34th PIANC World Congress 2018

Connecting Maritime Hubs Globally

Logistics Infrastructure

Panama City, Panama
7 - 11 May 2018

ISBN: 978-1-5108-8705-3

Printed from e-media with permission by:

Curran Associates, Inc.
57 Morehouse Lane
Red Hook, NY 12571

Some format issues inherent in the e-media version may also appear in this print version.

Printed by Curran Associates, Inc. (2019)

For permission requests, please contact
The World Association for Waterborne Transport Infrastructure (PIANC)
at the address below.

The World Association for Waterborne Transport Infrastructure (PIANC)
General Secretariat
Boulevard Roi Albert II, 20 – B. 3
B-1000 Brussels
Belgium

Phone: +32 2 553 71 61

info@pianc.org

Additional copies of this publication are available from:

Curran Associates, Inc.
57 Morehouse Lane
Red Hook, NY 12571 USA
Phone: 845-758-0400
Fax: 845-758-2633
Email: curran@proceedings.com
Web: www.proceedings.com

TABLE OF CONTENTS

COMMITTEES

INTERNATIONAL ORGANIZING COMMITTEE (IOC)

LOCAL ORGANIZING COMMITTEE (LOC)

Mr. Manuel Benítez, Congress Co-Chair (ACP)
Mr. Rogelio Gordón, Congress Co-Chair (ACP)
Mr. Oscar Vallarino, Local Executive Committee (ACP)
Ms. María Teresa Maure (ACP)
Mr. Ricardo Álvarez (ACP)

INTERNATIONAL SCIENTIFIC COMMITTEE (ISC)

Dr. Philippe Rigo, Vice President of the SC (Belgium)
Dr. Michael Fastenbauer, Chair (Austria)
Mr. Jean Michel Hiver (Belgium)
Mr. Jose Sanchez (USA)
Mr. Wu Peng, Chair (China)
Mr. Jim Stirling, Chair (UK)
Mr. Fabrice Daly, Chair (France)
Mr. Rik Thomas (Belgium)
Mr. Juergen Troegl, Chair (Austria)
Mr. Leonel Temer (Argentina)
Ms. Milou Wolters, Chair (Netherlands)
Mr. Stefan Devocht (Belgium)
Dr. Carsten Thorenz, Chair (Germany)
Dr. Tom De Mulder, Chair (Belgium)
Mr. Holger Schüttrumpf (Germany)
Mr. John Clarkson, Chair (USA)
Mr. Timothy Paulus, Chair (USA)
Dr. Jasna Muskatirovic (Serbia)
Mr. Ian White, Chair (UK)
Mr. Francisco Esteban Lefler, Chair (Spain)
Mr. Ignacio Sanchidrian Vidal, Chair (Spain)
Mr. Hartmut Bruehl, Chair (Germany)

Mr. H.N. Aswath (India)
Mr. Craig Philip, Chair (USA)
Mr. Edward Schmeltz (USA)
Mr. Hans Burcharth (Denmark)
Dr. Mohammadreza Allahyar (Iran)
Ms. Katrina Dodd (Australia)
Mr. Esteban Biondi, Chair (USA)
Mr. Pablo Arecco, Chair (Argentina)
Mr. Nicholas Pansic, Chair (USA)
Dr. Todd Bridges (USA)

LOCAL SCIENTIFIC COMMITTEE (LSC)

Dr. Dimas E. Portillo L., President of the SC (ACP)
Mr. Abdiel Perez (ACP)
Mr. Rigoberto Delgado (ACP)
Mr. Antonio Abrego (ACP)
Mr. Ricardo Álvarez (ACP)
Dr. Ricardo Ungo (ACP)
Mr. Juan Wong, Chair (Panama)

34th PIANC WORLD CONGRESS PANAMA 2018 May 7 to 11

CANAL DE PANAMÁ

COPYRIGHT AND REPRODUCTION

PREFACE

This book of abstracts contains 284 abstracts presented at the 34th PIANC World Congress held in Panama, Republic of Panama, from 7 to 11 May 2018. PIANC World Congress presents and discusses topics relevant to the waterborne transport and infrastructure sector. It rotates among PIANC's member countries every four years and it is open to members and non-members. Participants, have the opportunity to exchange knowledge and experiences.

PIANC's goal is to advance the sustainable development of shallow and deep-draft navigation issues including dredging and dredged material disposal, navigation and pot infrastructure, recreational navigation and related environmental matters.

The Book of Abstract is arranged according to the main Themes of the Congress namely: Inland Navigation, Dredging, Logistics & Infrastructure, Ports, Marinas, and Environment. Inside each Theme, the abstracts are arranged by the topics of the Theme.

Inland Navigation

- Waterway infrastructures: locks, weirs, riverbanks, others.
- Advances in navigation locks design after the Panama Canal Expansion
- Inland navigation channels: safety and reliability
- Integration of inland waterways into inter-modal supply chain
- Inland navigation, waterways, ports & terminals
- Systems and Infrastructures: design & management for inland navigation projects
- River Information Services (RIS, AIS, others)
- River navigation flow control

- Salt water intrusion
- Maintenance and operation of IW, transport & infrastructures
- Inland navigation in South America

Dredging (in the framework of port and navigation projects)

- Effective planning and execution of dredging projects
- Lessons learned from dredging projects worldwide
- Risk management in dredging projects
- Disposal site planning taking into account alternative land uses
- Current dredging & management innovations

Logistics & Infrastructure

- Challenges for capitalizing navigational channels bordering lands
- Synergies among airports, ports and inter-modal assets
- Integrated management of global supply chains
- Integrating ports and economic special Zones
- Provision of added value logistic services in maritime hubs
- Project management in the Panama Canal Expansion

Ports

- Impact and innovative solutions for hinterland ports
- Dry ports: better and more efficient
- Ports of the future: technologies, automation, traceability
- Port management models
- Maritime Port planning and operations
- Inter-modal connections
- Coastal and Port Engineering (in relation with navigation)
- Storm Surge & Tsunami barriers and Flood protection gates

Marinas

- Sustainable and resilient marina design-Marinas Working with Nature (best practices and case studies)
- Marinas as part of Tourism, Ports and Urban Master Plans (best practices and case studies)
- Planning Framework for Recreational & Tourism Navigation Infrastructure Systems (goals, strategies and challenges)
- Regulatory Framework and Private Marina Development in Countries in Transition (best practices and case studies)
- Regional Cooperation and Standardization of Recreational Navigation Information Systems Transition (best practices and case studies)
- Recreational components of mitigation plans for large navigation infrastructure projects

Environment

- Environmental management in navigation
- Climate change and emissions, energy efficiency, international regulations, carbon markets
- Fresh water availability for operations
- New technologies on infrastructure, pollution prevention, port reception facilities and ballast water
- Multiple purpose water resource systems (transport, energy, recreation, ecosystems, watersheds, potable water, ...)
- Navigation as a green transportation mode
- Societal awareness and responsibility, combining economic growth, environment / sustainability and welfare
- Climate Change Mitigation in the Maritime Industry

All these abstracts represent an invaluable source of information, knowledge, experience, and ideas useful for students, engineers, researchers, scientists, authorities and for everyone interested in the fields covered in this World Congress.

Name________________________________
Title________________________________
Position_____________________________
Affiliation___________________________

STUDY ON THE FUNCTIONING OF PORTS IN PRODUCTION AND LOGISTICS FOR EXPORT PROMOTION OF MARINE PRODUCTS

by

Masamitsu Nakaizumi[1] (Ph.D.), Shinpei Nagano[2], Takehito Horie[3] and Akira Nagano[4] (Ph.D.)

ABSTRACT

The Japanese Government is working with the private sector to double exports of domestic marine products. In this paper, we elucidate the current status regarding the functioning of fishing ports, seaports and airports in the export of marine products, and discuss ways to improve their production and logistics functions so as to promote exports. Fishing ports and seaports on the one hand, seaports and airports on the other, play important production and logistics functions from producers to consumers, as bases for the production of marine products, including hygiene management measures, and as bases for transportation, including measures to preserve freshness, respectively. Expanding high-value products, namely live, fresh and chilled products and value-added and ready-to-eat products such as fillets will be effective for expanding exports. We should be stimulating exports to East Asian countries using short international shipping routes traversed by ferries and RORO (roll-on, roll-off) ships, which offer fast loading and boarding for vehicles and freight from seaports in the producing regions, as well as exports all around the world by air from these regions, after strengthening measures to preserve freshness. It can be assumed that if seaports in the producing regions could offer efficient logistics functions that maintain a high level of freshness, this will lead to promotion of exports. It will also be important to link the production functions of fishing ports with a high level of hygiene management to the logistics functions of seaports.

1. INTRODUCTION

The Japanese government has teamed up with the private sector to double exports of marine products in an initiative to popularize Japanese culture and culinary tradition around the world. However, there has been a lack of research on the production and logistics required for these exports. Marine products differ greatly from other products in some respects, such as the measures that must be taken to manage hygiene (HACCP management, etc.), and the requirements for measures to preserve freshness in products can only be stored for very short periods. Such considerations would appear to have a substantial effect on the process of selecting the seaports and airports to be used for export, as well as means of transport.

The authors have analyzed existing statistical data and survey data, conducted our own site surveys, and conducted interviews with those concerned with the fishery and shipping industries to learn the current functioning of fishing ports, seaports, and airports with respect to conditions in the export of marine products. In this paper, we discuss how to improve the functioning of fishing ports and seaports so as to promote exports of marine products.

2. DETAILS OF RESEARCH

In this study, we have used FAO FishStat, the UN Comtrade Database and Japanese trade statistics to tabulate the exports of marine products (according to both volume and value) in terms of commodity items, the customs clearance locations, the export destination countries and means of transport. Also, using our analysis of data from commodity flow (movement) surveys of containerized freight and air freight, we have calculated the fractions of exports represented by the locations of production and in

[1] Professor, Tokyo University of Marine Science and Technology, Japan. mnakai0@kaiyodai.ac.jp

[2] Researcher, TS Visual Research Co., Ltd., Japan

[3] Team Leader, Alpha Hydraulic Engineering Consultants Co., Ltd., Japan

[4] President, Japan Construction Engineers' Association, Japan

terms of container types, as well as lots. The means of transport used, routes, lead times, costs, and forms of freight were estimated on the basis of materials collected from and interviews with producers, processors, shipping companies, customs offices and site surveys.

The above process was the basis for a discussion of the following items:

i. Current status in production and trade of marine products around the world and Japan
ii. Current status in the functioning of fishing ports, seaports, and airports vis-à-vis exports of marine products
iii. How to improve the functioning of fishing ports and seaports in order to promote exports of marine products

3. CURRENT STATUS IN PRODUCTION AND TRADE OF MARINE PRODUCTS AROUND THE WORLD AND JAPAN

3.1 World Trade in Marine Products

The worldwide annual per-person consumption of marine products is gradually but definitely increasing, sustained by increases in aquaculture production volumes all over the world (Figure 1) and exports and imports of marine products.

Per-person consumption of marine products

World 13.5kg(1990) → 19.0kg(2013) Japan 71.4kg(1990) → 48.6kg(2013)

Amounts for marine product trade

World (exports) US$55 billion(2000) →US$139 billion (2013)
World (imports) US$61 billion(2000) →US$135 billion (2013)

Considering these by region (Figure 2), we see that this is mainly within-region trade, in Europe and

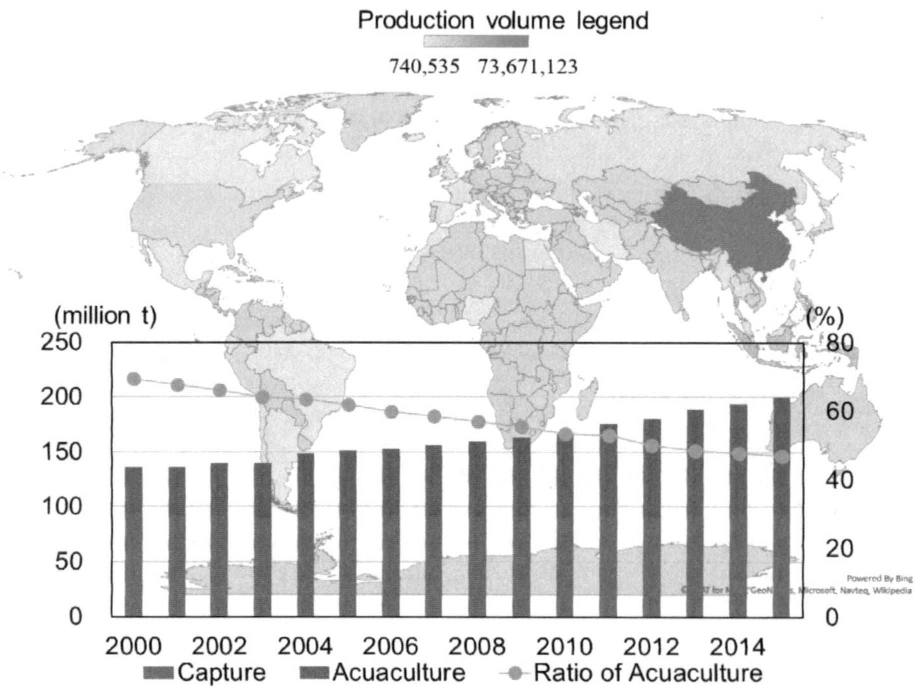

Source: FAO FishStat Online Query Panels

Figure 1: Trends in World Marine Production (Capture and Aquaculture)

Asia. Fresh and chilled products in particular are imported and exported between nations that are relatively close to each other. Trade within Europe makes up over 60% of the above totals.

Although the amount of annual per-person marine products consumption is tending downward among the Japanese public, Japan remains a key importer of marine products (#1 worldwide in 2000, #2 worldwide as of 2013). On the other hand, exports from Japan remain low despite efforts to promote exports for the purpose of the stability of domestic production prices resulting from the securing of new market.

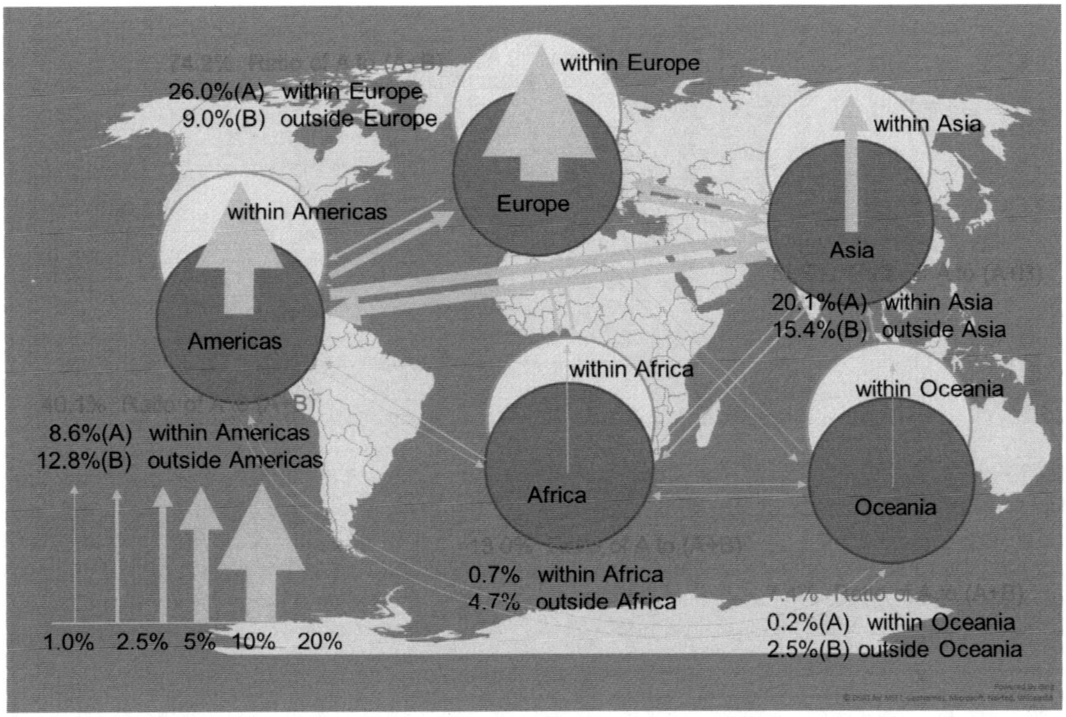

Source: UN Comtrade Database Database

Figure 2: Flows of Marine Product Trade Value by Region

3.2 Exports of Marine Products from Japan

Figure 3 presents the recent trends and current status in exports of marine products. Export figures plunged after the Great East Japan Earthquake in 2011, but have tended to increase in recent years, and have recovered to the levels predating the Lehman Brothers bankruptcy. If we break this down by export destination, we see that exports to South Korea, the EU and Russia have not returned to their peak in 2007, while they are increasing to new destinations such as Vietnam and Taiwan. Currently (as of 2014), Asian nations including Hong Kong, China, Thailand and Vietnam account for 77% of exports, and the USA, for 18%. In terms of species, exports are mostly scallops, pearls, mackerel, tuna, yellowtail and prepared foods, such as scallop adductor muscles, sea cucumber, and fish-paste products. Different nations import different items: pearls, scallop adductor muscles and dried sea cucumber are exported to Hong Kong; scallops and yellowtail to the USA; Scallops, Salmon and trout to China; and live fish and pollack to South Korea.

Due to fluctuations in domestic production, the value of the domestic production of scallops, mackerel, salmon, trout, farmed yellowtail, saury and other fish tends to vary widely. Therefore, they are exported in hopes of stabilizing the domestic prices. Exports of unshelled scallops (frozen) to China have grown; once those are processed, by shelling and placing in water, they are consumed domestically or exported to the USA. However, exports of scallops represent nearly 80% of Japanese domestic production and domestic consumption has fallen in half; hence, if we are to consider responding to the domestic consumption demand, it seems difficult going forward to increase exports at the current level of production. Salmon and trout are exported to China both dressed and semi-dressed; there, they are

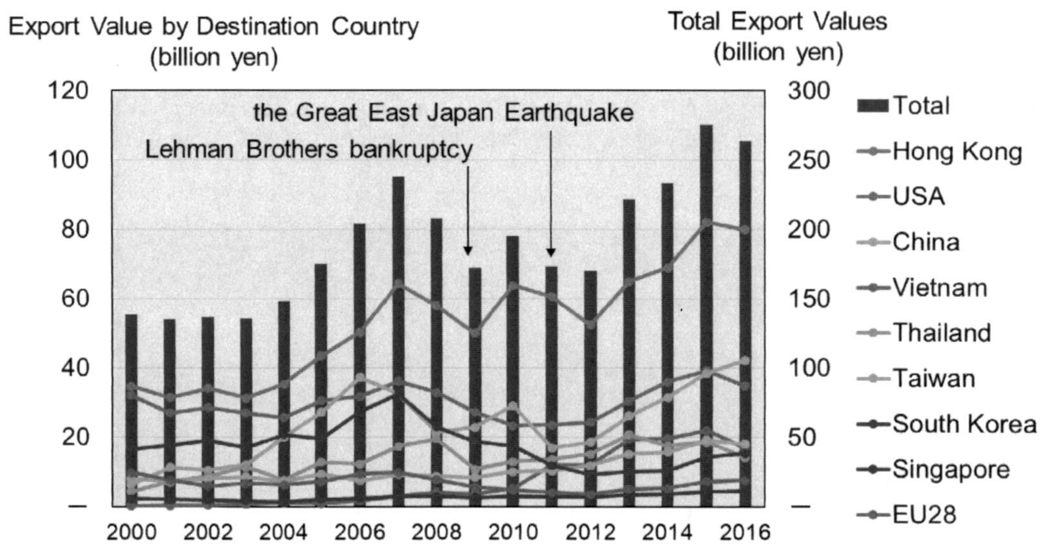

Source: Japanese Trade Statistics

Figure 3: Trends in Exports of Marine Products from Japan

filleted or otherwise processed and exported to Europe or the USA. It is feared that domestic production of these will fall in the future. There is a high demand for sea cucumber, which seems strong demand as a high-grade component of Chinese cuisine and fetches high prices; as such, nearly 80% of domestic production is directed to export and there are also fears for these resources. Farmed yellowtail, whose export began in 2008, are filleted or otherwise processed and exported to the USA. Bonito and tuna are exported to Thailand, where they are canned.

Species, Shipping Temperatures and Form of Products	2007			2014		
	Volume (t)	Value (million yen)	Average Price (yen/kg)	Volume (t)	Value (million yen)	Average Price (yen/kg)
All Types (=Total)	**562,632**	**133,836**	**238**	**422,167**	**133,671**	**317**
Live Fish	6,120	6,720	1,098	2,685	2,938	1,094
Fresh and Chilled Fish	43,401	17,845	411	9,147	7,056	771
Frozen Fish	461,725	64,479	140	327,433	51,358	157
Fillets (fresh and chilled)	1,738	2,742	1,578	1,438	3,130	2,177
Fillets (frozen)	9,025	9,749	1,080	7,878	10,494	1,332
Fish (dried, salted and smoked)	822	1,680	2,044	631	665	1,054
Crustaceans (frozen)	5,753	4,761	828	1,377	1,179	856
Crustaceans (non-frozen)	649	745	1,148	382	1,268	3,319
Shellfish, Cephalopods and Mollusks (live, fresh, and chilled)	8,259	2,427	294	9,198	3,549	386
Shellfish, Cephalopods and Mollusks (frozen and others)	25,142	22,689	902	58,749	47,779	813
Invertebrates (live, fresh and chilled)	Included in the group of Shellfish, Cephalopods and Mollusks			1,638	1,075	656
Invertebrates (dried and others)				1,615	3,183	1,971

Source: Japanese Trade Statistics

Table 1: Comparison of Export Volume and Value and their Average Prices by Shipping Temperature and Form of Product (Japan)

Table 1 provides a comparison between the peak year (2007) and 2014 for yen-denominated exports of marine products, divided into live, fresh, frozen and round, and filleted. The value of the lowest-priced products (frozen) dropped from 50% to 40%, and the value of the high-priced products (fresh and chilled) fell by about one-half; meanwhile, the comparatively expensive mollusks and cephalopods (shellfish, squid, sea squirt, etc.) increased in export value by about 40%, thanks to the rise in exports of scallops.

Exports contribute to the stability of domestic production prices. Nevertheless, authorities must be alert for issues including future effects on the available resources, depending on fish species, and the hollowing of the domestic processing industry.

4. CURRENT STATUS IN THE FUNCTIONING OF FISHING PORTS, SEAPORTS AND AIRPORTS VIS-À-VIS EXPORTS OF MARINE PRODUCTS

4.1 Customs Clearance Locations for Export of Marine Products

Figure 4 shows the values of exports at the customs clearance locations (2014). The customs clearance locations can be regarded as generally either seaports or airports of origin. In terms of maritime freight, the amounts for the ports of Tokyo, Kawasaki and Yokohama and the ports of Osaka and Kobe, which serve a large number of regular international shipping routes to ports and harbors all over the world and are designated as international strategic ports, stand out from other locations. Following them are the ports of Tomakomai, Shimonoseki and Hakata, designated as international base ports that serve the key producing regions of Hokkaido (northern Japan), Chugoku and Kyushu (western Japan), respectively.

Shimonoseki and Hakata are superior locations for their proximity to the important export destination countries of South Korea and China. For air freight, Narita International Airport, Haneda International Airport and Kansai International Airport similarly serve many air routes connecting to destinations all over the world. New Chitose Airport and Fukuoka International Airport lie in key producing regions and many exports pass through them as well.

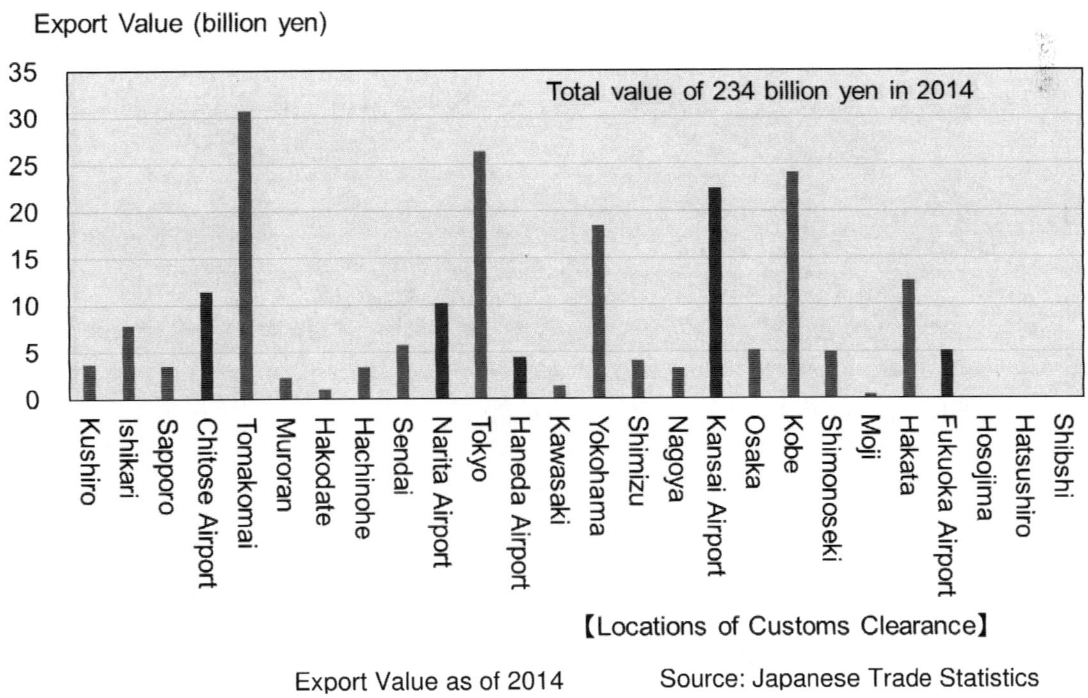

Figure 4: Value of Marine Products Exported through Customs Clearance Locations (Seaports and Airports in Japan)

4.2 Measures to Manage Hygiene in Fishing Ports and Processing Plants

Much has been done to manage hygiene, particularly in production base fishing ports, with the objective of increasing domestic consumption and promoting exports of Japanese marine products. This has taken the form of both hardware and procedures. Facilities have been upgraded, including sheds erected on wharfs and equipped with clean sea water supply facilities, closed-structure sorting and auctioning hall, etc., while regular testing and inspections and other soft countermeasures are conducted. Since production base fishing ports handle a large amount of fish, have emphasized facilities capable of maintaining high-level hygiene. Over 60% of production base fishing ports now have such facilities.

Export of marine products requires registration and certification of the processing facilities that handle products for export, as required by the destination countries. Notably, the strict Hazard Analysis and Critical Point (HACCP, i.e. hygiene management) standards are required for exports to the EU and the USA. Most of the certificated facilities for export to the EU and registered facilities for export to the USA are located in or near production base fishing ports.

4.3 Transport Means Used, by Fish Species, Shipping Temperature, and Form of Product

Figure 5 shows the means of transport used for export, and the marine products by fish species, shipping temperature, and form of products, for each transport means.

Containerized freight maritime transport is the most commonly used means for carrying freight out of the seaports of origin, but in terms of value, air freight transport accounts for nearly 30% of the total. Containerized freight maritime transport is used for marine products (frozen) and prepared foods, non-containerized freight maritime transport (ferries, RORO ships, live fish carriers, etc.) is used for live fish and other marine products (fresh, chilled, frozen), and air freight transport is used for live fish, other marine products (fresh, chilled, fillets), prepared foods, pearls, ornamental fish and others.

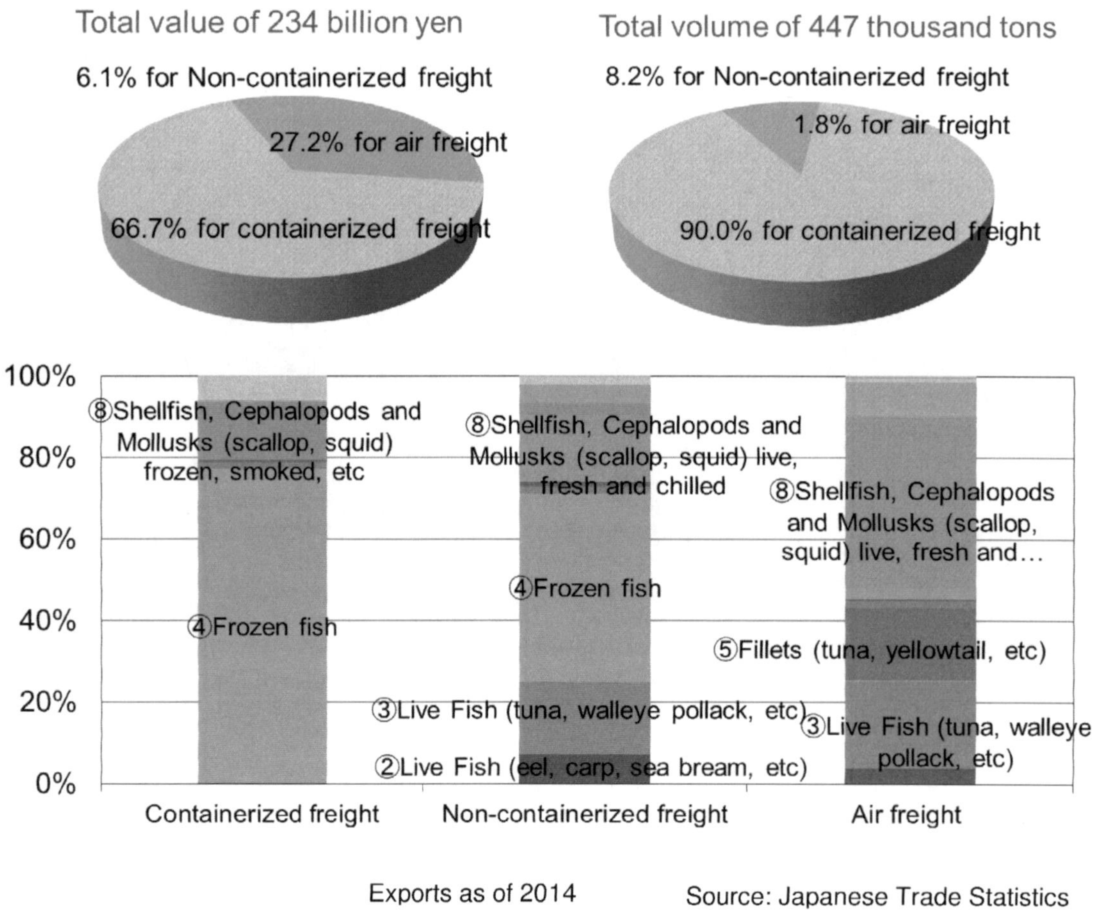

Figure 5: Means of Transport, Classified by Species, Shipping Temperature and Form of Product

Appropriate management of temperatures over a low range, including frozen, is critical for marine products. Thus, reefer containers account for over 90% of the containers used. Fresh or chilled marine products are packed into styrofoam boxes with cold storage and refrigerating agent in other maritime freight, or into refrigerated trucks on ferries or RORO ships. They are also packed into Styrofoam boxes for air transport. Of exports shipped in consolidated form, 99% of maritime containers are FCL freight, and 92% of air freight is less than load (LCL).

The unit price of marine product freight – classified by species, shipping temperature, and form of product – was calculated (as of 2014) and plotted, by means of transport, with the former on the vertical axis and the latter on the horizontal axis, in Figure 6. The marine products exported were also subdivided by the means of transport. Air freight transport was employed for expensive items and for carrying items quickly that must remain fresh, such as live fish, fresh products and chilled products. Containerized freight maritime transport was used for frozen, dried and prepared foods.

When relatively "medium- to low-priced" products, live, fresh and chilled, were sent by non-containerized freight maritime transport in export to neighboring countries, ferries and RORO ships (carrying live fish transport vehicles, refrigerated trucks, or palletized freight placed in the hold) were used for nearby routes, as well as live fish carriers, fresh and frozen fish carriers.

Pearls, coral and ornamental fish are excluded

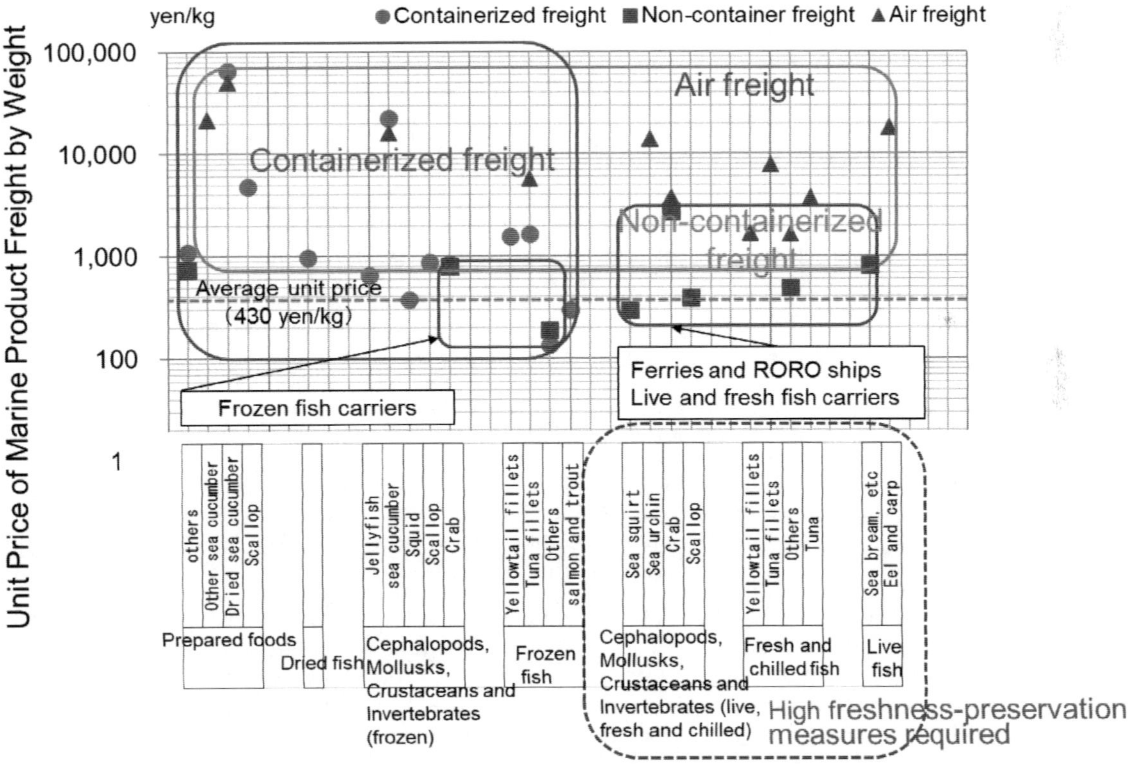

Figures as of 2014 Source: Japanese Trade Statistics and others

Figure 6: Unit Price of Marine Product Freight, Classified by Species, Shipping Temperature and Form of Product

4.4 Characteristics of Different Means of Transport

The freight transportation costs by weight for each means of transport for marine products (as of 2016), the travel time for maritime transport and the flight time for air transport were estimated on the basis of materials received from shipping companies and airlines, site surveys and interviews with those concerned. The costs were plotted for each means of transport on the vertical axis versus the times plotted on the horizontal axis of Figure 7.

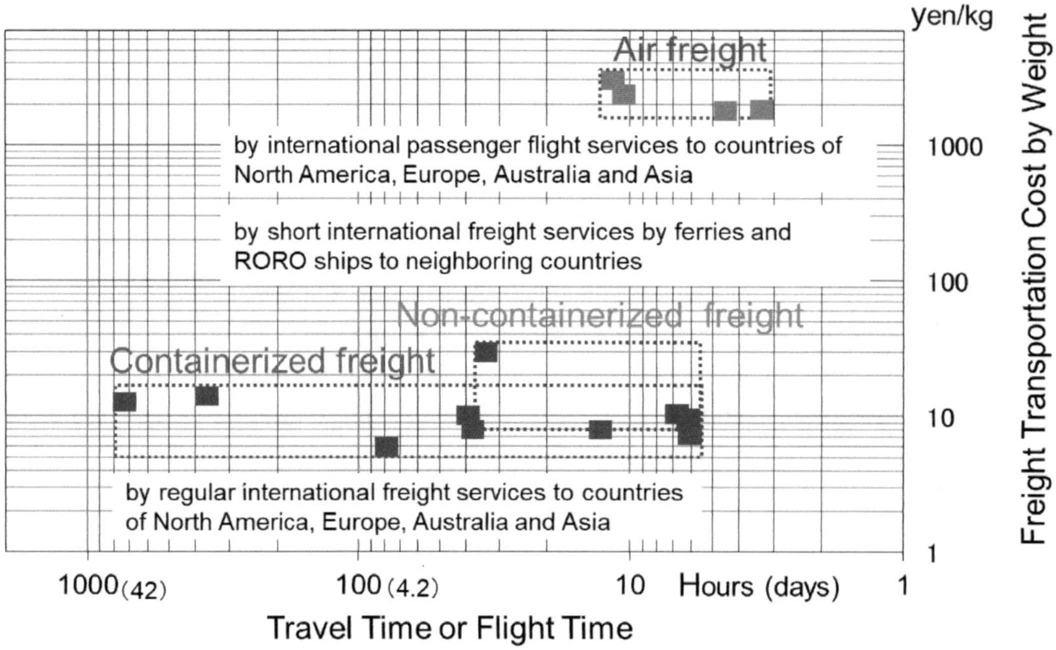

Figure 7: Freight Transportation Cost by Weight, and Travel or Flight Time

The freight transport conditions were estimated as follows.

[Freight transport conditions]

Air freight (consolidated, direct-flight freight): From New Chitose/Narita/Kansai/Fukuoka to China/Hong Kong/North America/Europe/Australia

Maritime freight (containerized): From Tokyo/Yokohama/Osaka/Hanshin/Hakata, by 20-ft reefer container to South Korea/China/Hong Kong/North America/Australia

Maritime freight (non-containerized): From Hakata or Shimonoseki by ferry (25 t refrigerated truck or live fish transport vehicle on board) to South Korea or China

In the figure, we will see that containerized freight maritime transport is low-cost and the travel time is long, from 6 hours to about one month; in contrast, air freight transport is expensive, but the flight time is short, from 3.5 to 13 hours. Non-containerized freight maritime transport offers comparatively medium to low costs, and travel time is relatively short, 6 hours to about 1.5 days.

Turning to the size of the lots transported, the average weight of freight carried by the various services was calculated using data from sources. This was 98 freight tons for containerized freight maritime transport and 380 kg for air. Air freight lots are regarded as much smaller than the large lots in containerized freight maritime transport. As for non-containerized freight maritime transport, the surveys and interviews with those concerned indicated that for refrigerated trucks and live fish transport vehicles, the main vehicles were large, at about 25-t size.

Thus, the aforementioned results show that containerized freight maritime transport offers long travel times but low transportation costs and the capability to handle large lots; air freight transport has high costs and can handle only small lots, but allows short travel times. Non-containerized freight maritime transport has features that one could call intermediate between containerized freight and air freight transports.

4.5 Means of Transport Used for Export of Live, fresh and Chilled Marine Products

Table 2 summarizes the means of transport employed for exports of live, fresh and chilled marine products (2014), according to trade statistics. This shows that live fish were exported to the nearby countries of South Korea, Hong Kong and Taiwan, while very few went to China. This was because the formalities for obtaining the export certificates necessary for exporting live fish had not been adequately

Species, Shipping Temperature, Products Form	Export Destination Country	Export Value (million yen)	Means of Transport Used
Live Fish	South Korea	1,710	Sea breams transported from Shimonoseki, Hakata or Izuhara Port by ferry (carrying live fish transport vehicles), or from Uwajima Port by live fish carriers
	China	6	Live fish transported from Takamatsu or Uwajima Port by live fish carrier
	Hong Kong	129	Eels transported from Chubu or Fukuoka Airport
	Taiwan	497	Eels transported from Chubu or Fukuoka Airport
Fresh and Chilled Fish	South Korea	1,861	Pollack, sea bream, etc. transported from Shimonoseki Port by ferry (carrying refrigerated trucks, palletized freight placed in the hold, or containerized freight)
	China	174	Tuna transported from Fukuoka Airport
	Hong Kong	796	Fish transported from Fukuoka or Haneda Airport
	USA	911	Fish transported from Fukuoka, Haneda or Narita Airport
	Guam	1,686	Directly unloading tuna captured overseas at port in Guam
Fillets (fresh and chilled)	Hong Kong	310	Tuna transported from Haneda Airport
	USA	1,499	Yellowtail transported from Fukuoka Airport
	Canada	147	Yellowtail transported from Fukuoka Airport
Crustaceans (non-frozen)	South Korea	324	Crab transported from Shimonoseki or Hakata Port by ferry (carrying palletized freight placed in the hold)
	Taiwan	412	Crab transported from New Chitose Airport
	Hong Kong	110	Crab transported from New Chitose Airport
Shellfish, Cephalopods and Mollusks (live, fresh, and chilled)	South Korea	1,914	Live unshelled scallops transported from Shimonoseki or Hakata Port by ferry (carrying live fish transport vehicles or palletized freight placed in the hold)
	China	498	Live unshelled scallops transported from New Chitose Airport or from Shimonoseki or Hakata Port by ferry or RORO ship (carrying live fish transport vehicles or palletized freight placed in the hold)
	Hong Kong	703	Live unshelled scallops transported from New Chitose Airport
	Taiwan	199	Live unshelled scallops transported from New Chitose or Narita Airport
Invertebrates (live, fresh and chilled)	Hong Kong	241	Sea urchin transported from Narita Airport
	South Korea	444	Squirt transported from Shimonoseki or Hakata Port by ferry (carrying live fish transport vehicles or palletized freight placed in the hold)

Figures as of 2014

Table 2: Means of Transport Used for Exporting Live, Fresh and Chilled Marine Products

completed (the procedure outline was established in July 2014). Fish (fresh and chilled) saw a dramatic drop in South Korea, which could be attributed to the economic situation.

If we leave out the live, fresh and chilled marine products via air freight transport, we see that exports to South Korea (Busan) from the ports of Shimonoseki, Hakata, Izuhara, Uwajima and other ports make up a large part of exports, as shown in red in Table 2; exports to China account for but little. The means of maritime transport were ferries and RORO ships (carrying refrigerated trucks and live fish transport vehicles on board) from the ports of Shimonoseki and Hakata and live fish carriers and other carriers from Uwajima Port and other seaports.

Table 3 shows the frequency of service provided on regular short international shipping routes (ferries and RORO ships). For services from Shimonoseki, Hakata and Izuhara to South Korea and China, short shipping routes to these countries (shown in red in the table) are used. The reasons for this appear to be the fact that ferries and RORO ships can carry live fish transport vehicles and refrigerated trucks on board, and that since loading deadlines for vehicles and freight are closer to departure times and are faster for vehicles and freight, and the loading process is quick, this suits the need to maintain a high level of freshness during transportation.

	Shipping Company	Ports of Call	Type of Ship	Frequency per Week	Travel Time
Japan-South Korea	Sunstar Line Co., Ltd.	Busan-Tokyo-Yokohama-Osaka-Busan-Osaka	RORO Container loading	1 from Tokyo 2 from Osaka	Tokyo→Busan 41hrs
	Sunstar Line Co., Ltd.	Osaka-Busan-Osaka	Ferry (RORO) Container loading	3	Osaka→Busan 19hrs
	Kampu Ferry Co., Ltd.	Shimonoseki-Busan-Shimonoseki	Ferry Container loading	7 (every day)	Shimonoseki→Busan 12hrs
	Camellia Line Co., Ltd.	Hakata-Busan-Hakata	Ferry (RORO) Container loading	7 (every day)	Hakata→Busan 5.5hrs
	Sunstar Line Co., Ltd.	Busan-Tsuruga-Kanazawa-Masan-Busan	RORO Container loading	1	Kanazawa→Masan 20hrs
		Busan-Tsuruga-Kanazawa-Busan-Masan-Busan	RORO Container loading	1	Kanazawa→Busan 24hrs
	Daea Express Shipping	Izuhara-Busan-Hidakatsu-Busan-Izuhara	Ferry	7 (every day)	Izuhara→Busan 2hrs Hidakatsu→Busan 1hrs
Japan-China	Shanghai Ferry Co., Ltd.	Osaka-Shanghai-Osaka	Ferry Container loading	1	Osaka→Shanghai 46.5hrs
	Japan-China International Ferry Co., Ltd.	Osaka-Shanghai-Kobe-Shanghai-Osaka	Ferry Container loading	0.5 (every two weeks)	Osaka→Shanghai 47hrs
	Shanghai Super Express Co., Ltd.	Hakata-Shanghai-Hakata （2016.1以降休止）	RORO (Ferry) Container loading	2	Hakata→Shanghai 29hrs
	Orient Ferry Ltd.	Shimonoseki-Qingdao-Shimonoseki (canceled since2016)	Ferry Container loading	2	Shimonoseki→Qingdao 29hrs
	Suzhou Shimonoseki Ferry Co., Ltd.	Shimonoseki-Taicang-Shimonoseki	RORO Container loading	2	Shimonoseki→Taicang 33hrs
Japan-South Korea-Russia	DBS Cruise Ferry Co., Ltd.	Donghai-Vladivostok-Donghai-Sakai-Donghai-Vladivostok	Ferry Container loading	1	Sakai→Donghai 14hrs Donghai→Vladivostok 20hrs Sakai→Vladivostok 41hrs

Source: Materials from Shipping Companies

Table 3: Regular Short International Shipping Routes (Ferries and RORO Ships)

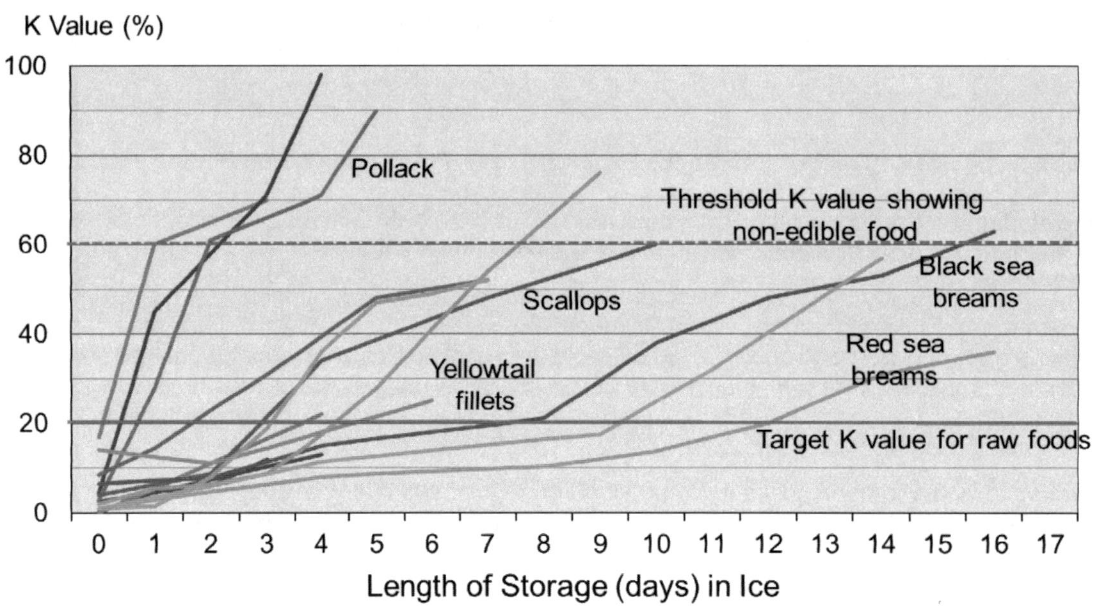

Note: scallops stored in 5℃

Figure 8: Spoilage of Marine Products versus Length of Storage (days) in Ice

4.6 Spoilage, Lead Times and Expansion of Use of Transportation Means

The "K value" indicates the level of decomposition of ATP for fish and is an index of the freshness of marine products. The lower the ATP decomposition, the lower the K value, and the better the freshness of the fish. Figure 8 provides a summary of the existing research findings on the change in K value that shows the falling freshness of marine products stored in ice. It was not clear how the fish were treated just after they expired or how they were stored in ice, but different species exhibit varying initial periods of freshness and speeds of spoilage.

The target K value for raw foods such as sashimi is 20% or less, so for export of fresh or chilled marine products, the lead time (defined in this paper as the time necessary to reach from producers to consumers) must be shorter than the period for which the marine products remain fresh. For example, previous research results have reported that bleeding the fish out or keeping it in slurry sea ice immediately after capture or harvest can delay spoilage. It is also assumed possible to delay spoilage of marine products during export, by the method of storing fish in ice immediately after capture or harvest and there onward.

Interviews with those concerned about export of live, fresh and chilled marine products yielded many concerns about the death of live fish and spoilage of fresh and chilled marine products, due to the time required to obtain the export certificates. It is assumed that streamlining the release of these certificates can shorten the lead time and increase the use of the means of transport.

4.7 Means and Routes of Transport of Marine Products from Production to Export

The fisheries and aquaculture statistics and trade statistics were analysed on the basis of the means of transport used and the relationship between the customs locations and the seaports and airports used. This revealed that the routes from producers to consumers and the means of transport used. Figure 9 presents the routes from the producers to export, using scallops as an example.

Scallops caught and harvested at nearby producing grounds are unloaded at production base fishing ports and packed for export in registered and certificated processing facilities located behind the fishing

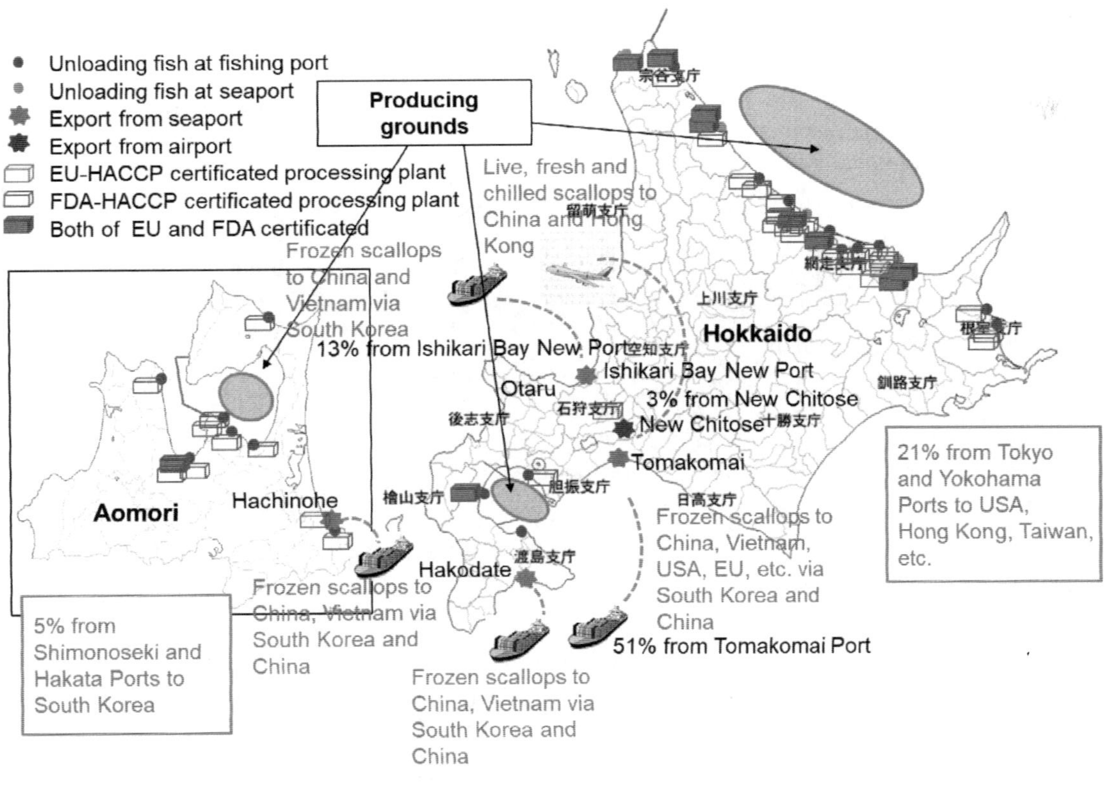

Figure 9: Means and Routes of Transport of Scallops from Production to Export

ports. Frozen scallops are exported from the ports of Tomakomai (51% in terms of value), Ishikari Bay New Port (13%), Tokyo, and Yokohama (21%). Live, fresh and chilled scallops are exported via New Chitose Airport to China and Hong Kong and from the ports of Shimonoseki and Hakata to South Korea.

4.8 Conclusions

We see from the preceding analysis that with regard to the export of marine products, fishing ports, seaports and airports on the one hand, and seaports and airports on the other, play important production and logistics functions from producers to consumers, as bases for the production of marine products, including hygiene management measures, and as bases for transportation, including measures to preserve freshness, respectively.

Expanding high-value products, namely live, fresh and chilled marine products, and value-added and ready-to-eat products such as fillets will be effective for expanding exports. We should be stimulating exports to East Asian countries using short international shipping routes traversed by ferries and RORO ships, which offer fast loading and boarding for vehicles and freight from seaports in the producing regions, as well as exports all around the world by air from these areas, after strengthening measures to preserve freshness.

5. HOW TO IMPROVE THE FUNCTIONING OF FISHING PORTS AND SEAPORTS IN ORDER TO PROMOTE EXPORT OF MARINE PRODUCTS

5.1 Current Exports from Hokkaido of Northern Japan

The amount of exports from seaports and airports in Hokkaido was ¥61.3 billion (26.2% of the total for Japan; all figures here for 2014). The amounts from the principal ports and airports were Tomakomai Port, ¥31.9 billion, Ishikari Bay New Port, ¥9.6 billion, and New Chitose Airport, ¥11.5 billion. The total exports of live, fresh and chilled marine products produced in Hokkaido was ¥4.9 billion; these amounted to ¥1.8 billion exported to Taiwan, Hong Kong and other destinations from New Chitose Airport. Exports to South Korea and China were mostly carried to Shimonoseki Port or Hakata Port (¥3.0 billion), overshadowing those exported from Ishikari Bay New Port and other ports in Hokkaido (¥160 million).

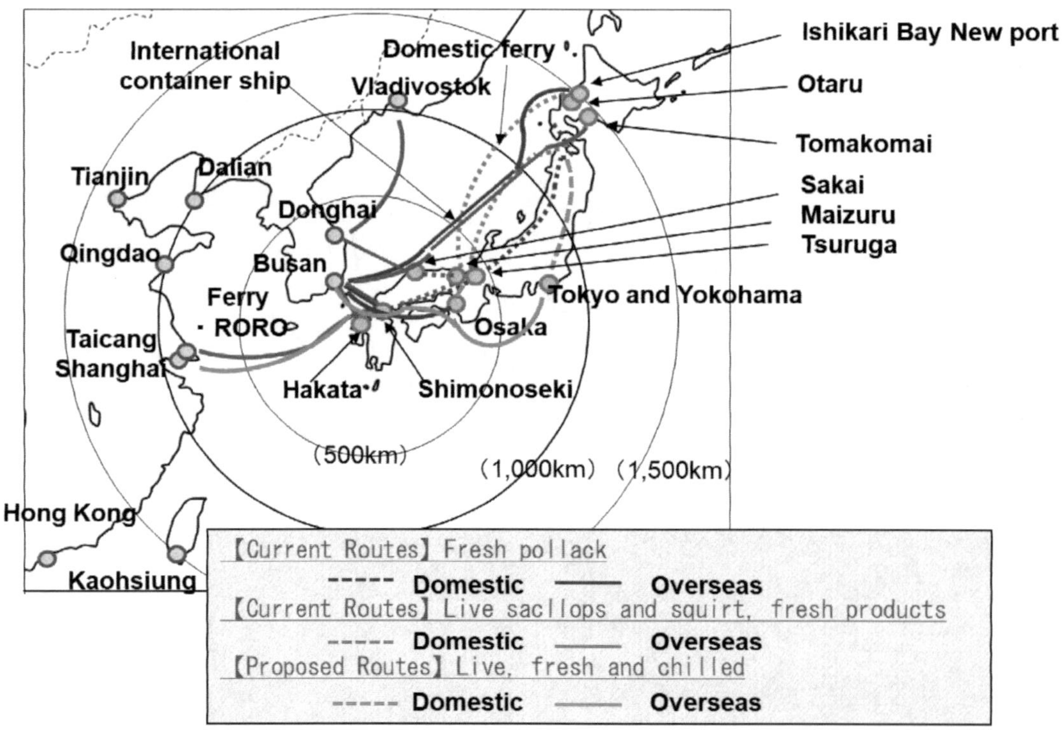

Figure 10: Means of Transport and Routes Used from Hokkaido to South Korea and China

	Species and Shipping Temperature		Means of Transport and Routes	Export Port	Export Destination Country	Export value (million yen) as of 2014	Form of Freight	Transportation Cost (thousand yen)	Lead Time (days to export)	Current Use	Potential for Expansion of Use
Current (as of 2014) Routes	Fresh Pollack	A	Producing regions → by refrigerated trucks → Shimonoseki Port → by ferry carrying refrigerated trucks or palletized freight placed in the hold (daily services) → Busan Port	Non-Hokkaido (Shimonoseki Port)	South Korea (Busan Port)	472	Refrigerated truck	810	4~4.5 (3)	High	
						131	Containerized freight	690	4~4.5 (3)	High	
		B	Producing region → by refrigerated trucks → Osaka Port → by ferry or RORO ship carrying reefer containerized freight (5services a week) → Busan Port	Non-Hokkaido (Osaka Port)	South Korea (Busan Port)	37	Containerized freight	530	4.5~5 (2)	Low	High
		C	Producing region → by refrigerated trucks → Ishikari Bay New Port → by full-container ship carrying reefer containerized freight (once a week) → Busan Port	Hokkaido (Ishikari Bay New Port)	South Korea (Busan Port)	25	Containerized freight	260	4~4.5 (1.5)※	Low	High
	Live Scallops and Live Squirt	D	Producing regions → by live fish transport vehicles → Otaru Port → by domestic ferry carrying live fish transport vehicles (daily service) → Tsuruga or Maizuru Port → by live fish transport vehicles → Shimonoseki or Hakata Port → by ferry carrying live fish transport vehicles (daily service) → Busan Port	Non-Hokkaido (Shimonoseki or Hakata Port)	South Korea (Busan Port)	2,277	Live fish transport vehicle	1,000	3~3.5 (2)	High	
		E	Producing regions → by live fish transport vehicles → Otaru Port → by domestic ferry carrying live fish transport vehicles (daily service) → Tsuruga or Maizuru Port → by live fish transport vehicles → Sakai Port → by ferry carrying live fish transport vehicles (once a week) → Donghai Port (Busan Port → Shimonoseki Port, on return way)	Non-Hokkaido (Sakai Port)	South Kore (Donghai Port)	89	Live fish transport vehicle	1,050	3~3.5 (2)	Low	
		F	Producing regions → by live fish transport vehicles → Otaru Port → by domestic ferry carrying live fish transport vehicles (daily service) → Tsuruga or Maizuru Port → by live fish transport vehicles → Shimonoseki Port → by ferry carrying live fish transport vehicles (twice a week) → Taicang Port	Non-Hokkaido (Shimonoseki Port)	China (Taicang Port)	116	Live fish transport vehicle	1,550	4~4.5 (2)	Low	High
	Fresh Marine Products (others)	G	Producing regions → Tomakomai Port → by full-container ship carrying reefer containerized freight (twice a week) → Busan Port	Hokkaido (Tomakomai Port)	South Korea (Bussan Port)	11	Containerized freight	260	4~4.5 (1.5)※	Low	High
Proposed Routes	Live, Fresh and Chilled Marine Products	H	Producing regions → Tokyo or Yokohama Port → by RORO ship carrying live fish transport vehicles, refrigerated trucks or reefer containerized freight (once a week) → (Osaka Port) → Busan Port	Non-Hokkaido (Tokyo or Yokohama Port)	South Korea (Busan Port)		Live fish transport vehicle	1,920	4~4.5 (1.5)		No possibility due to high transportation costs and long lead time
							Refrigerated truck	1,500	4~4.5 (1.5)		
							Containerized freight	500	5~5.5 (1.5)※		
		I	Producing regions → Osaka Port → by ferry carrying live fish transport vehicles, refrigerated trucks or reefer containerized freight (once or twice a week) → Shanghai Port	Non-Hokkaido (Osaka Port)	China (Shanghai Port)		Live fish transport vehicle	1,680	3.5~4 (2)		
							Refrigerated truck	1,360	3.5~4 (2)		Possible
							Containerized freight	530	4.5~5 (2)		possible

Table 4: Currently Used Means of Transport and Routes for Export of Live, Fresh and Chilled Marine Products Produced in Hokkaido to South Korea and China, and New Proposals

5.2 Potential for Expansion of Exports from Seaports in Hokkaido

The goal is to expand exports from seaports in Hokkaido, but currently, nearly all exports from Hokkaido leave from Shimonoseki Port or Hakata Port to South Korea or China, so it is important for us to clarify and examine the reasons for this. We examined materials from shipping companies, conducted a survey and interviewed those concerned about the current (as of 2014) status with regard to means of transport and routes used by all live, fresh and chilled marine products originating in Hokkaido, including exports from seaports outside Hokkaido, and about new proposals. Based on that, we estimated the transportation costs and lead times, the results of which are in Table 4.

The conditions for transporting freight assumed for the estimates are as follows. The means of transport and routes are shown in Figure 10.

[Freight transport conditions]

> Actual weight 7 t (corresponding to volume of approximately 25 m^3) of marine products, or styrofoam boxes iced with refrigerating agent (box volume 0.034 m^3), 700 boxes in a 25-t live fish transport vehicle, in a refrigerated truck or in a 20-ft reefer container

The lead times marked a ※ in the table are of concern. When the issuance of export certificates requires 2-3 days, this lengthens the lead time by 0.5 - 1.5 days, and may impact the cost of maintaining freshness and transportation.

The means of transport and routes mostly used for shipping to South Korea at present are A and D, through Shimonoseki Port or Hakata Port, non-Hokkaido. The transportation costs for these are higher than for other means of transport and routes, but the lead times are around 3 - 4.5 days. Since carriers (live fish transport vehicles and refrigerated trucks) leave the ports of Shimonoseki and Hakata by these regular shipping routes every day, so one could assume that not only are these reasonable from an economic point of view, but they also allow a stable, dependable means of tranport. C and G are direct export routes from Ishikari Bay New Port and Tomakomai Port and are less costly, but they offer only the scant 1 and 2 services per week, respectively. Therefore, they are only used for a fraction of shipments. Very few shipments for China have been sent by route F through Shimonoseki Port, non-Hokkaido; route F costs 1.5 as much as route D, presumably the reason for the low use of this option.

The following are necessary conditions for expanding exports of live, fresh and chilled marine products from seaports in Hokkaido to South Korea:

i) A continuing loosening of import restrictions and further streamlining of the issuance of export certificates (both by national and local administrative agencies);
ii) Development and introduction of technologies related to export by container and more advanced methods for preserving freshness (by producers, exporters and port/harbor operators).

These would be expected to render feasible the use in earnest of full-container ships from Hokkaido. It is over 2,500 km from seaports in Hokkaido to China, requiring 1 - 1.5 days more than when shipping live, fresh and chilled marine products to South Korea; it is not realistic to look to these routes. So, presumably for China, while advocating for i) and ii) above, more exports to China should be sent via the ferry from Shimonoseki Port (F), and a ferry (I) from Osaka Port would be possible to use. A feasibility study on expanding or new uses of modes of transport and routes should be conducted going forward.

5.3 Current Exports from Kyushu, Chugoku and Shikoku of Western Japan

The amount of exports from seaports and airports in Kyushu, Chugoku and Shikoku are ¥29.5 billion (12.6% of the total for Japan; all figures here for 2014). The amounts from the principal seaports and airports were Hakata Port, ¥12.5 billion; Shimonoseki Port, ¥4.7 billion, and Fukuoka Airport, ¥5.0 billion. Live, fresh and chilled marine products are exported from Fukuoka Airport to the USA, China, Hong Kong, and Taiwan. A look at the trends in export figures (Figure 11) from the principal seaports of origin reveals that exports from Shimonoseki Port are at a low level compared with its former peak. On the other hand, Hakata Port has seen a growth trend despite the Lehman Brothers Shock and the 2011 Great East Japan Earthquake; it currently accounts for some 7% of all of Japan's exports in yen terms.

In terms of destination countries, the main export destinations from Shimonoseki Port (Figure 12) were South Korea (¥4.0 billion) and China (¥400 million); the great part of these (¥3.9 billion) were live, fresh and chilled marine products. About 50% of all exports were products of Kyushu, Chugoku and Shikoku, while about 40% were from Hokkaido. The main destinations from Hakata Port (Figure 13) were South Korea (¥2.1 billion), China (¥600 million) and the USA (¥6.2 billion). Most of the products sent to South Korea and China were live, fresh and chilled marine products, and about 80% of these were from Hokkaido.

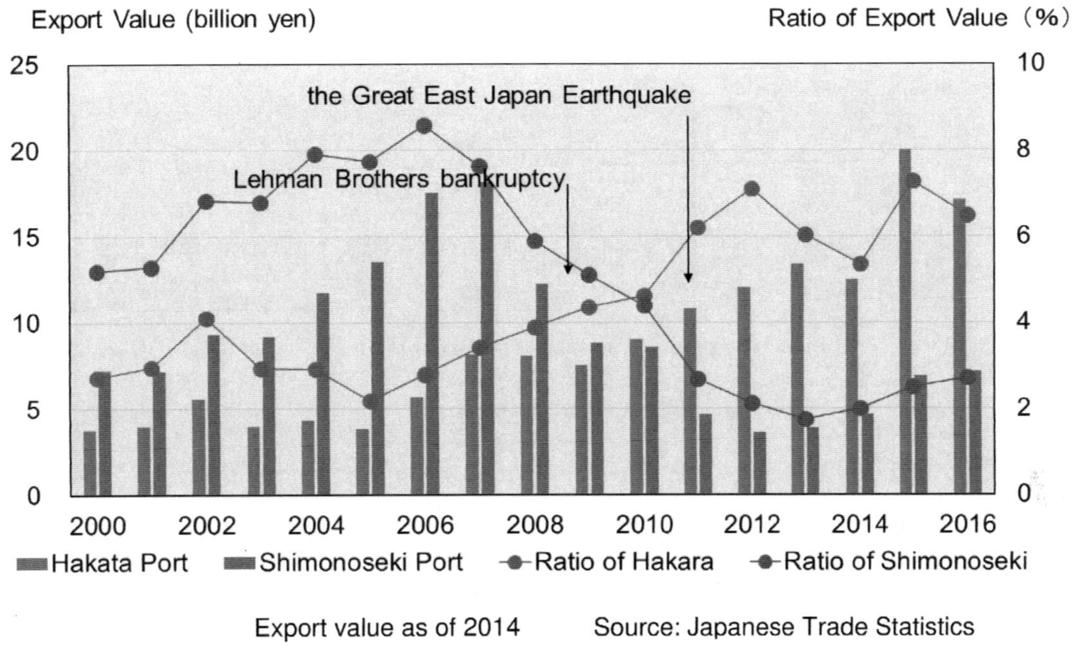

Figure 11: Trends in Yen-Denominated Exports from Shimonoseki and Hakata Ports

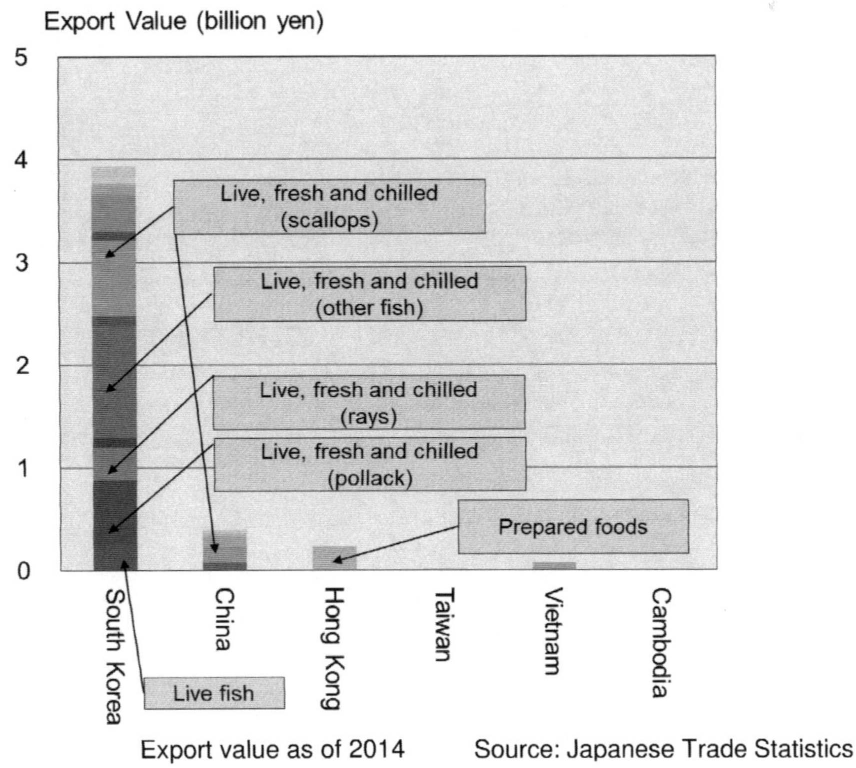

Figure 12: Yen-Denominated Exports from Shimonoseki Port, by Destination and Product Type

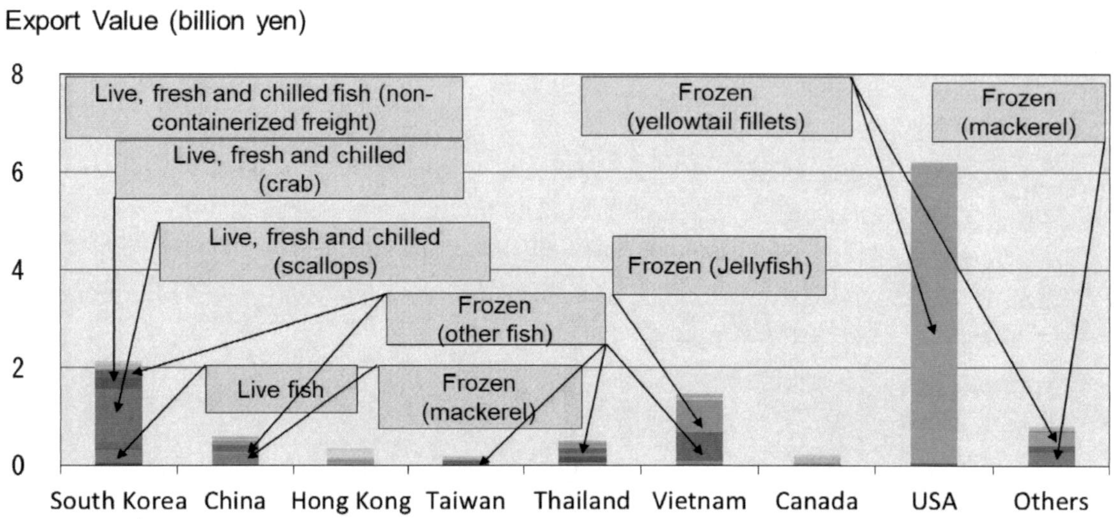

Export value as of 2014 Source: Japanese Trade Statistics

Figure 13: Yen-Denominated Exports from Hakata Port, by Destination and Product Type

In addition to serving regular international containerized freight routes, Hakata Port (Figure 14) provides international RORO service to Busan and Kaohsiung. There are also bases for marine, land and air freight modes gathered within a radius of 5 km. Its regular international shipping routes, that take advantage of its proximity to East Asia, along with its cooperation with multimodal transport, are superior in terms of high-speed and regularity of service.

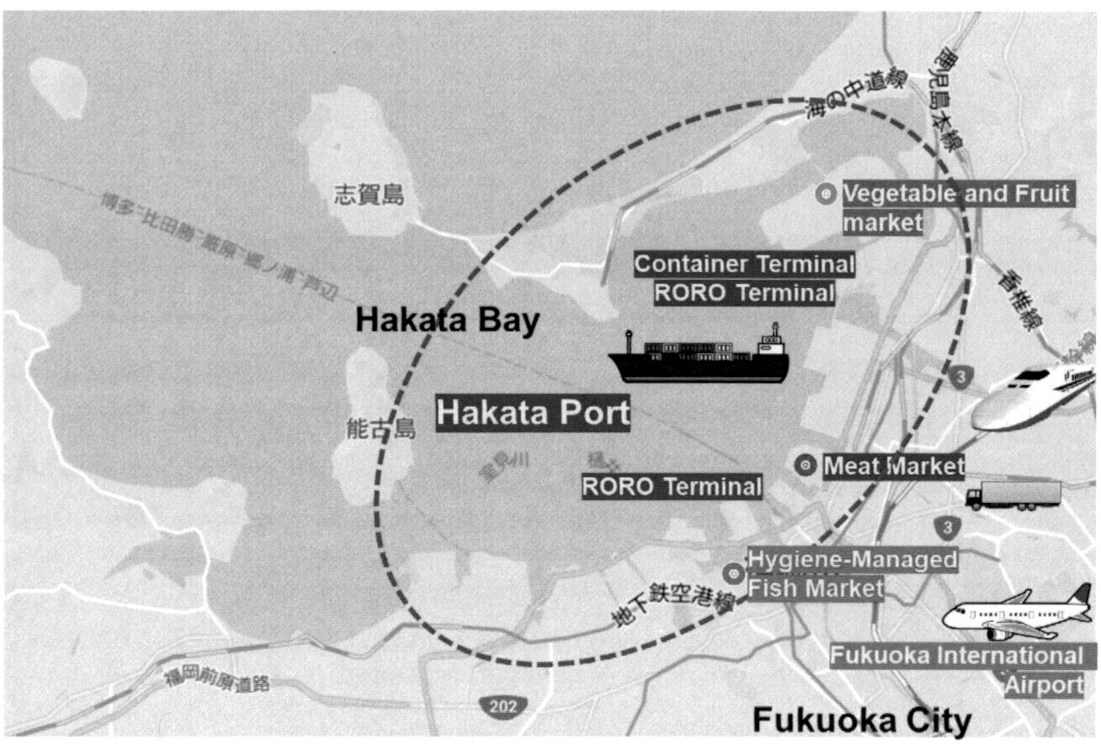

Source: Google Map

Figure 14: Production and Logistics Functions Available at Hakata

5.4 Potential for Expansion of Exports from Seaports in Kyushu, Chugoku and Shikoku

With a wide set of services, including fish, agricultural produce and meat markets, Hakata, Fukuoka City, has established an export-oriented environment. However, even though a large volume of fresh fish come to Hakata Fishing Port by sea and by land from Kyushu, Chugoku and Shikoku, only a tiny portion of them are exported from Hakata Port. It will be the most effective way to export more marine products unloaded at the fishing ports in or adjoining the seaports like Hakata Port for the purpose of the stability and expansion of exports from these seaports, by reducing costs and time freight pickup and ensuring that freshness can be preserved. However, interviews with those concerned pointed out a breakdown of information necessary for addressing export at every stage from production to export.

The key advantages of the ports of Shimonoseki and Hakata are their proximity to South Korea, China and domestic producing regions, the speedy short shipping routes to export destination countries and their well-coordinated inspection system. In addition, Hakata Port has well-established regular international freight routes with Southeast Asia, Europe and North America. It has a strong connection with South Korea; nearly all its live, fresh and chilled marine products for export are bound for South Korea.

Regrettably, the South Korean economy is continuing to falter and South Korea continues to enforce restrictions on imports of marine products from Japan; these are impediments to expanding exports. As for exports to China, although an agreement was reached in July, 2014 on handling of certification for exports of live marine products, and the export environment was improved, partly because the Chinese apparel industry has moved its bases to Southeast Asia, the Shimonoseki – Qingdao ferry and the Hakata – Shanghai RORO services have been canceled since 2016 (Table 3). These were great setbacks to expansion of exports to China.

The necessary issues for increasing exports of live, fresh and chilled products from the ports of Shimonoseki and Hakata are the following: i) A continuing loosening of Korean and Chinese import restrictions and further streamlining of the issuance of export certification; and ii) Development and introduction of technologies related to export by container and more methods for preserving a high level of freshness. Addressing those issues can be expected to encourage use of ferries and RORO ships and enable both widespread use of full-container ships and the expansion of exports to China.

5.5 Conclusions

Following the ports of Tokyo, Kawasaki and Yokohama and the ports of Osaka and Kobe, the most important international strategic ports in Japan, as transportation bases for exporting marine products, are Tomakomai Port in Hokkaido, and the ports of Hakata and Shimonoseki in Kyushu, Chugoku and Shikoku (located in the key producing regions), which are the international base ports. The ports of Shimonoseki and Hakata have the advantage as they are the closest to export destination countries.

Depending on the characteristics of the means of transport (travel times, costs, lot size, etc.) used for exporting live and fresh marine products, it is the means of transport and the domestic and overseas routes that dictate whether or not seaports have the logistics function that can transport marine products economically and reliably within a lead time that can preserve the required freshness.

In order to expand exports of live and fresh marine products from producers, it will be necessary i) to loosen Korean and Chinese import restrictions and to streamline the issuance of export certificates; and ii) to develop and introduce transportation technologies for containerized freight and measures for for preserving a high level of freshness from the stage of fish catching or harvesting onward. Taking action on those issues will presumably promote use of ferries and RORO ships and enable widespread use of full-container ships for exports. It will also improve the functioning for promoting exports from the seaports.

It appears to be important for the production functions of fishing ports with high-level hygiene management and the logistics functions of seaports serving as transportation bases to work for expanding exports in collaboration. By sharing information between producers, exporters, fishing ports and seaport authorities—namely information on the live and fresh fish collected in a market, and about the demand at the destinations and the capabilities of means of transport—the logistics functions of seaports, on which the reliable and efficient export of live and fresh fish depends, will improve further.

6. IN CLOSING

Fishing ports and seaports on the one hand, and seaports and airports on the other, play important production and logistics functions from producers to consumers, as bases for the production of marine products, including hygiene management measures, and as bases for transportation, including measures to preserve freshness, respectively.

Expanding high-value products, namely live, fresh and chilled marine products, and value-added and ready-to-eat products such as fillets will be effective for expanding exports. We should be stimulating exports to East Asian countries using short shipping routes traversed by ferries and RORO ships, which offer fast loading and boarding for vehicles and freight from seaports in the producing regions, as well as exports all around the world by air from these regions, after strengthening measures to preserve freshness.

Depending on the characteristics of the means of transport (travel times, costs, lot size, etc.) used for exporting live and fresh marine products, it is the means of transport and the domestic and overseas routes that dictate whether or not seaports have the logistics functions that can transport marine products economically and reliably within a lead time that can preserve the required freshness.

The following points must be addressed in order to expand exports of live, fresh and chilled marine products from Japanese producers:

(1) Promoting improvement of fishing ports as production bases that deal with high-level hygiene management
(2) Firm linkage of the production functions of fishing ports with the logistics functions of seaports
(3) Loosening of import restrictions and streamlining of the issuance for export certificates
(4) Development and introduction of transportation technologies for containerized freight and measures for preserving a high level of freshness from the stage of fish catching or harvesting onward

References

A. Kataishi et al. (2016). Study on Freshness-Management Issues of Marine products and Measures for Preserving Freshness, Proceedings of Japanese Fisheries Engineering.

E. Okazaki et al. (2007). Evauation of Freshness and Taste of Marine Products, Research Journal of Food and Agriculture 30(6).

FAO FishStatJ: http://www.fao.org/fishery/statistics/software/fishstatj/en

Fisheries and Forestry Department of Hokkaid (2007). Manual for Preserving Freshenss of Fresh Marine Products.

Finance Ministry of Japan: Trade Statistics: http://www.customs.go.jp/toukei/srch/index.htm

M. Nakaizumi et al. (2008). Economic Impact of Quality Control and Hygiene Management of Fishery Products in Fishing Ports in Hokkaido, Proceedings of Civil Engineering in the Ocean, Vol.24.

Ministry of Land, Infrastructure, Transport and Tourism (2014). Commodity Flow Survey of Containerized Freight.

Ministry of Land, Infrastructure, Transport and Tourism (2014). Commodity Flow Survey of Air Freight.

M. Kimura (2003). Study on Prserving Freshness of Scallop Adductor Muscle.

UN Comtrade Database: https://comtrade.un.org/

Y. Hiraoka (2007). Quality and Labeling of Farmed Fish and their Quality Management during Distribution.

DESIGN OF THE STRUCTURAL HEALTH MONITORING SYSTEM FOR THE THIRD BRIDGE OVER THE PANAMA CANAL

Gloribel Arlin Cespedes Melendez

ABSTRACT

A signature bridge is an important part of the nation's infrastructure because of the function it serves, but also because its attests to a country's economic strength and technological advancement. Panama, is building a major cable-stayed bridge with a design life of in excess of one hundreds of years so it would be beneficial to monitor it so that any departure from assumptions made during design are detected early. During its service life, circumstances and conditions may change, resulting in different types and magnitudes of live load, material variations, natural disasters, and human factors. To ensure safety, the operational functionality and durability of the bridge, it is important to have a comprehensive understanding of the reliability of its structural components. Therefore, a Structural Health Monitoring (SHM) system is needed. This paper outlines a proposed SHM plan for what will be the world's largest cable-stayed bridge with a concrete superstructure – The Third Bridge over the Panama Canal at the Atlantic Side. The proposed bridge, which has a 530 m main span, is located in Colon, Panama, one of the most corrosive environments in the world. The paper summarizes the overall objectives, the design principles, recommended types and locations of sensors. The proposed SHM system will include the following aspects: sensor arrays, a data acquisition and transmission system, data processing and control, a health diagnosis methodology, early warning alarms, and a security assessment process. An important focus of the system will relate specifically to durability and the monitoring of corrosion in reinforced concrete. Data collected will support and help optimize decision-making on future maintenance and repair. Tension force monitoring in cable-anchorage system will be proposed as well to detect changes that might occur due to corrosion or fatigue.

INTRODUCTION

Transportation holds the economic lifeline of a nation. The construction and maintenance of bridges is essential for a nation's infrastructure, and symbolizes the economy development and technology improvement of the nation as well.

Cable-stayed bridges are magnificent long-span bridges that have been extensively constructed throughout the world. They have been chosen due to their rigidity, good aerodynamics, good seismic resistance, aesthetics, and in most of the cases, due to their constructability. Avoiding the disruption of maritime operations, which is the case in the Panama Canal. Within the last century, two cable-stayed bridges have been constructed over the Panama Canal: the Centennial Bridge (420 m), and the Third Bridge at the Atlantic Side (530 m), and there are plans to build a new one in the Pacific Side (+/-500 m) that will carry a light rail.

Structural health monitoring (SHM) systems can be a useful tool to ensure safety, serviceability, durability, and sustainability of structures. By permanently installing a number of sensors, continuously measuring relevant parameters, we can record the real loads, environmental conditions, real behaviors, and real evolution process and performance of a structure during its service time. Anomalies and/or deterioration or damage that may induce adverse effects on service or safety can be detected and evaluated through these systems. SHM help us to understand the behavior or a real, full-scale structure under synthesized loads and environmental conditions. Furthermore, these systems and technologies allow a better understanding of the influence of natural disasters, which are hard to simulate in small-scale models or in an accelerated durability test in the laboratory. SHM is a new safety and management tool that ideally complements traditional methods like visual inspection and modelling. Monitoring even allows a better planning of the inspection and maintenance activities, shifting from scheduled interventions to on-demand inspection and maintenance [1].

THIRD BRIDGE OVER THE PANAMA CANAL AT THE ATLANTIC SIDE

The Third Bridge over the Panama Canal at the Atlantic side consists of a dual carriageway Cable-Stayed Bridge with a vertical clearance of 75 m (246 ft), a total length of 1050 m (3445 ft) and a main span of 530 m (1739 ft). Each carriageway consists of two 3.6 m (11.8 ft) width lanes, two 1.8 m (6ft) external shoulders and two 0.6 m (2ft) internal shoulders to provide satisfactory capacity and emergency for commercial vehicles.

The main cable stayed bridge comprises five spans with twin towers with an upper Delta shape. The cable supported superstructure is in the form of continuous box girder. The cable stayed bridge span arrangement is 79m + 181m + 530m + 181m + 79m (Fig. 1).

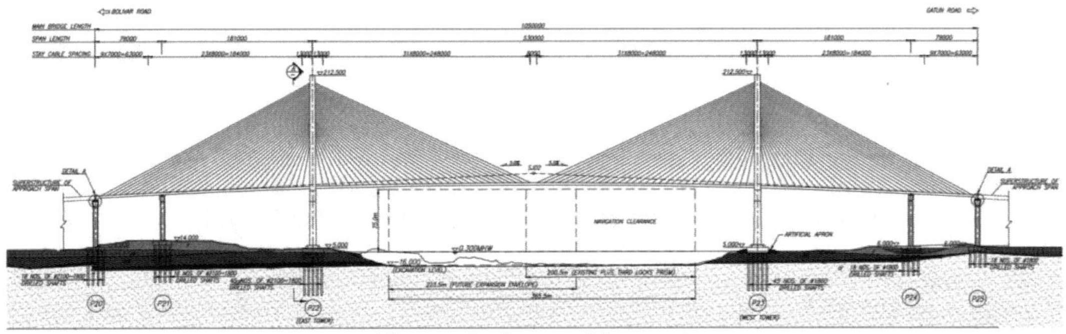

Figure 1: General Layout

The deck superstructure is composed by the cast-in-situ concrete box girder. The total width of bridge deck is 23.6m and the depth of deck girder structure (edge) is 2.6m, as shown in Fig. 2. It is supported by stay cables radiating out from the two towers. A diaphragm is proposed at every 7-8m with rib width of 220~300mm. The prestressed concrete main girders have 33 units of cast in-situ segments on each side of tower and are being constructed by balanced cantilever method. The bridge was designed following AASHTO Specifications [2] to resist strong-motion seismic impact (10% probability of exceedance in 50 years), vessel collision and wind loads.

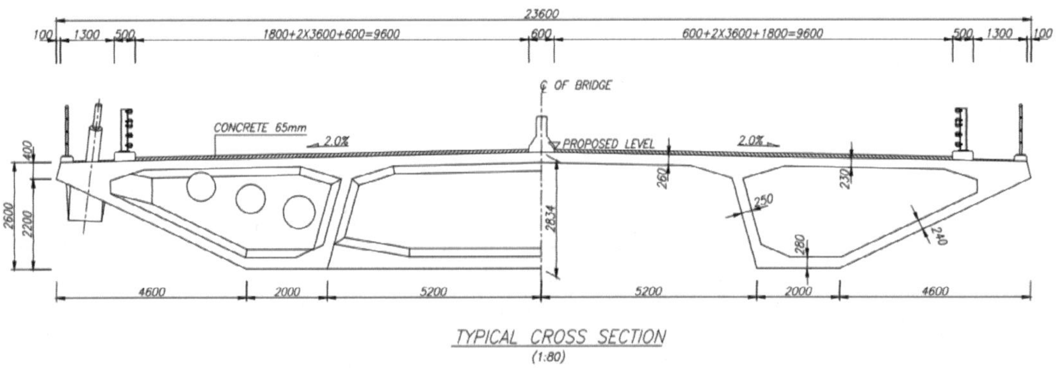

Figure 2: Typical section of deck

DESIGNING AN SHM SYSTEM

When designing a monitoring system, it is fundamentally important to design as an integrated system, which has all data flowing to a single database and presented through a single user interface. The integration between the different sensing technologies can be simultaneously installed on the structure, for example, fiber optic sensors, vibrating wire sensors, weather stations and corrosion sensors, can be achieved at several levels. Different sensors can be connected to the same data-logger: otherwise, several data-loggers can report to a single data management system, which is generally a personal computer installed either on site or at a remote location. Even though many of the sensors and systems provide their own software for data management and presentation, one of the goals of this design would be to provide a single integrated interface that does not require the end user to learn and interact with different user interfaces.

There are several functional requirements that must be considered when designing SHM system:

a. The system must implement advanced techniques, good performance, long term stability, and economic value rationality.
b. The system must have the capability to transmit data, process view, archive document and share long distance information.
c. The system must be able to collect data synchronously, real-time, long term, and hierarchically.
d. The system must have capability to assess, control, and calibrates itself.
e. The system must be able to identify damages and evaluate structural health.

f. The system must be reusable and upgradable [3].

Having the potential responses and locations, the goal was to select the sensors that have the appropriate specifications to sense the expected responses and that are appropriate for installation in the specific environmental conditions and under the technical constraints found in the structure.
Each monitoring project presents its peculiarities and although it is possible to standardize most elements of a monitoring system, each application is unique in the way they are combined. It is however possible to classify the monitoring components according to several categories. A Structural Health Monitoring (SHM) System is composed of the following modules:

1. Sensory System
2. Data acquisition and transmission system
3. Data processing and control system
4. Structural health evaluation system
5. Structural health data management system, and
6. Inspection and maintenance system [4]

Modules 1 and 2 are sensors, data loggers and cabling networks for signal collection, processing and transmission; Modules 3, 4 and 5 are computer systems assigned with different functions of system control and system operation. Module 6 has the purpose of performing the inspection and minor maintenance works for modules 1 and 2. This modular architecture approach has been used in the Sutong Bridge [5] and the Donghai Crossing.

SENSORY SYSTEM

Sensors provide the basic data for a structural health detection and monitoring system. Their performance directly determines ultimate success of the detection and monitoring methods. Sensors need to record the structural mechanical conditions, and directly transform the measured parameters like strain, displacement, acceleration into signals for output. They must sense the variation of external environmental conditions, which needs sufficient reliability, sensibility and high reaction velocity to reflect the external information promptly and accurately [6, 7]. Usually the selection of sensors takes the following factors into consideration: the type, precision, resolution, frequency response and dynamic range; the distribution positions, the influence degree of the surrounding dynamic environment and measurement noise.

The physical response occurring in bridges are measured by current, voltage, resistance, magnetic flux, acoustic and optical fibers. The objectives of a sensory system are the following [4]:

(a) To measure the local and global levels of responses;
(b) To integrate the measured data with analyzed data for correlation analysis and model updating;
(c) To acquire data in a consistent manner and retrievable manner for subsequent diagnostic and prognostic analysis of structural health and damage;
(d) Use contemporary commercially available sensors;
(e) To carry out supplementary measurements by removable/portable sensors for validation/updating of analytical models
(f) To carry out cross-calibration of different types of sensors at, at least, one typical location.

For the design of sensor modules and sensing technologies for cable-stayed bridges, three aspects should be considered: the variable type, the sensor type, and the positioning of the installed sensors. For the selection of sensory systems, the following criteria should be taken into account: operational bandwidth; magnitude and frequency response over that bandwidth; sensitivity and accuracy; power supply requirements; physical characteristics (dimensions, weight and material); environmental operating conditions such as temperature ranges; type of output signal; any signal conditioning requirements; and costs.

NUMBER AND LOCATION FOR PERMANENT MONITORING

The selection of locations to monitor depends on the configuration of the bridge, structural loading, structural materials and environmental loading. A summary of the sensors is:

WEATHER AND TEMPERATURE LOADS SENSORS

Optical fiber Bragg grating temperature sensors will be mounted on the box girder, and at the pylons. These sensors are measuring temperatures, which are correlated with the strain gauge measurements. Three (3) weather stations (ultrasonic anemoscopes [8], hygrometer, barometers and rain fall gauges) measuring wind speed, wind direction, air humidity and air temperature. Two (2) will be mounted on the top of the pylons, the other at road level. The wind measurements serve as a reference for the stay cable vibrations. The air humidity and temperature complete the meteorological information.

CORROSION SENSORS

For the cable-stayed main span, the major concrete structural components are pylons, intermediate piers and pier shafts. As the pylons are more important than the intermediate piers, one pylon is selected for monitoring. Three plane sections of the pile cap, the water-saturated zone, splash zone, and dry zone will be monitored because they represent micro-environments. Each section will be instrumented with four (4) sets of corrosion monitoring sensors at one plane section. The anchorage zone will be instrumented with ten (10) sets of corrosions. As a result, a total set of eleven (11) corrosions for each pylon will be installed on the pylons. For the superstructure, the cross-section of the box girder will be selected for permanent monitoring [9].

HIGHWAY TRAFFIC SENSORS

For measuring the weight and speed of vehicles, weigh-in-motion sensors will be used. Four (4) stations will be embedded underground in pavement. These weighbridge sensing components are optical fiber Bragg grating strain sensors that are made in steel-concrete composite structure, which have the same life with bridge itself.

ACCELEROMETERS AND GLOBAL POSITIONING SYSTEM (GPS)

Since vibration of cables is a common phenomenon, forty four (44) triaxial force balance accelerometers will be mounted on the stay cables [10]. Twenty-one (21) biaxial piezoelectric accelerometers will be installed on the deck and twenty (20) on the pylons. Six (6) GPS receivers will be installed on the top of both pylons, the middle of the main span, both ends of the main bridge, and the bank nearby the bridge, respectively. The receiver installed on the bank will be referred as the base station [11].

DEFORMATION SENSORS AND TILTMETERS

For deformation monitoring of the girder and pylon, several permanent stations around the bridge should be used during bridge construction, which will be utilized for the current monitoring task, so that all data can be later unified. There will be a total of 14 monitoring points the deck, four (4) on the piers and ten biaxial tiltmeter (10) [12] on the pylons [13].

STRAIN SENSORS

These sensors are mainly needed to observe torsions due to heavy wind and traffic. Twelve (12) optical fiber Bragg grating strain sensors will be mounted on 3 steel outriggers of the cables, one on each side. Two will be mounted on the rail level in the concrete and four will be mounted on the lower side of the bridge.

Table 1. Monitoring sensors summary

Monitoring Item	Sensor	Quantity
Weather monitoring	Ultrasonic-type anemometers	3
	Barometers	3
	Rainfall gauges	3
	Hygrometers	3
Temperature loads monitoring	Optical fiber Bragg grating temperature sensors	28
Corrosion status monitoring	Corrosion Sensors	22
Highway traffic loads monitoring	Dynamic weigh-in-motion stations	4
	Digital video cameras	2
Dynamic features monitoring	Fixed and portable servo-type accelerometers	41
Cable forces monitoring	Portable servo-type accelerometers	44

Geometry monitoring	GPS	6
	Displacement traducers	4
	Tiltmeters	10
Strain/Stress monitoring	Dynamic strain gauges	28
	Static strain gauges	12
Fatigue life monitoring	Dynamic strain gauges	12
Articulation monitoring	Displacement transducers	2
	Total	227

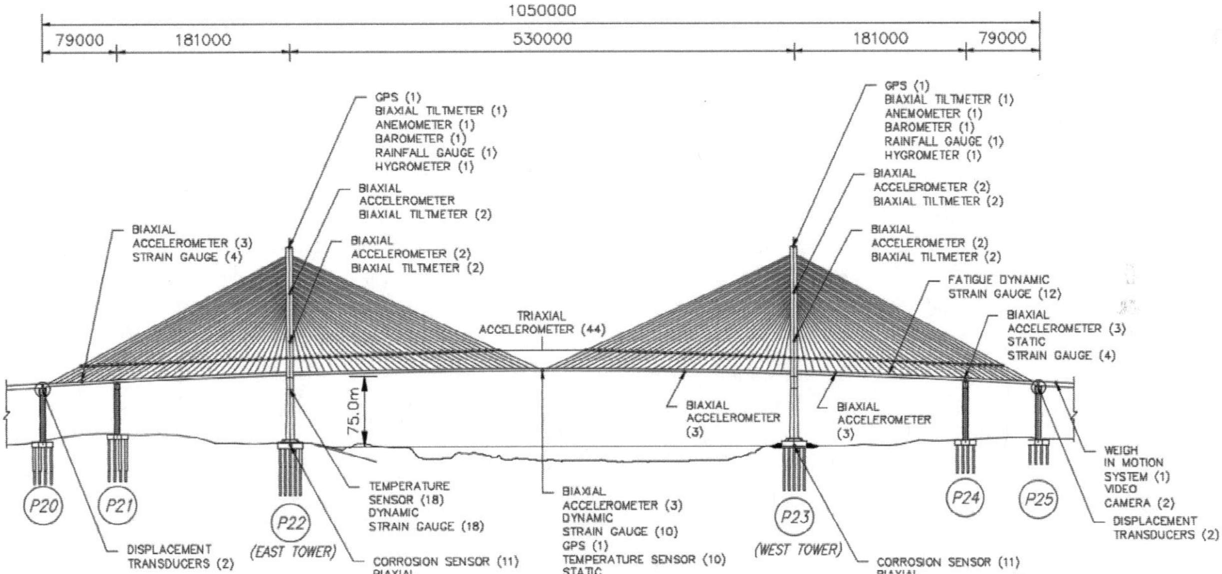

Fig. 3. Layout of Sensor System

CONCLUSIONS

This paper presents design criteria for a structural health monitoring system for the Third Bridge over the Panama Canal at the Atlantic Side, which is a cable stayed bridge. It offers a reference and guide to design a health monitoring system.

Generally, SHM system consists of six modules sensory system, data acquisition and transmission system, data processing and control system, structural health evaluation system, structural health data management system, and inspection and maintenance system. Two aspects of the design of sensor are presented: location and type of sensor. Scheme, hardware, software and data acquisition strategies are needed to be designed for data acquisition module. Technology, scheme, instrument and software data transmission are to be properly chosen. The speed and distance are two critical factors impacting scheme and technology of data transmission system.

Further research will be done in the concept of SHM by periodic Non-Destructive Evaluation (NDE) surveys since they have been used as a tool for structural maintenance and asset management.

REFERENCES

[1] D. Inaudi, "Integrated Structural Health Monitoring Systems for Bridges," 2009.

[2] L. AASHTO, "AASHTO LRFD Bridge Design Specifications, Customary U.S. Units, 7th Edition, with 2015 and 2016 Interim Revisions," ed: American Association of State Highway and Transportation Officials, Washington, DC, 2014.

[3] S. A. Putra, B. Riyanto, A. Harsoyo, and A. Imam, "Conceptual design of multi-agent system for suramadu bridge structural health monitoring system," *Telkomnika (Telecommunication Computing Electronics and Control),* vol. 13, pp. 1079-1088, 2015.

[4] K. Wong and Y. Ni, "Structural health monitoring of cable-supported bridges in Hong Kong," ed: Woodhead Pub., 2009.

[5] X. Dong, Y. Zhang, H. Xu, and Y. Ni, "Research and design of structural health monitoring system for the Sutong Bridge," 2005.

[6] W. Xiao-Guang and X. Zu-En, "Development of long-span bridge health monitoring," *Journal of Chang'an University (Natural Science Edition),* vol. 23, pp. 39-42, 2003.

[7] H. Dajian and X. Jun, "State of arts of health monitor ing techniques for long span br idges," *Br idge Construe—tion,* vol. 6, pp. 69-73, 2002.

[8] H. Li, "Design approach of health monitoring system for cable-stayed bridges," *Integrative Oncology: Principles and Practice,* p. 307, 2005.

[9] X. G. Hua, Y. Q. Ni, H. F. Zhou, and J. M. Ko, "Instrumentation for durability monitoring of a long-span cable-stayed bridge," in *Smart Structures and Materials 2005: Sensors and Smart Structures Technologies for Civil, Mechanical, and Aerospace Systems, 7 March 2005*, USA, 2005, pp. 982-91.

[10] B. Peeters, G. Couvreur, O. Razinkov, C. Kündig, H. Van Der Auweraer, and G. De Roeck, "Continuous monitoring of the Øresund Bridge: system and data analysis," *Structures and Infrastructure Engineering,* vol. 5, pp. 395-405, 2009.

[11] L. Hui and O. Jinping, "Full implementations of structural health monitoring systems for long-span bridges and large-span domes," in *Nondestructive Evaluation and Health Monitoring of Aerospace Materials, Composites, and Civil Infrastructure V, 28 Feb. 2006*, USA, 2006, pp. 61761-1.

[12] H. Wang, A. Li, J. Niu, Z. Zong, and J. Li, "Long-term monitoring of wind characteristics at Sutong Bridge site," *Journal of Wind Engineering and Industrial Aerodynamics,* vol. 115, pp. 39-47, 2013.

[13] Z. Guanyong, J. a.-m. c. Zhijun Zhang, and C. Fischer, "Structural Health Monitoring of a Long-span Cable-stayed Bridge," *Journal of Intelligent Material Systems & Structures,* vol. 18, pp. 835-843, 08// 2007.

Panama within the Silk Road

Napoleon once said "China is a sleeping giant, let it sleep, because when it wakes up the world will move" and certainly China has been the great surprise of the 20th century and the 21st century. Its high rates of economic growth, tied to the common interest of improving the welfare of its citizens, have strengthened the country, placing it as the second global economy. China has reduced the level of poverty and increased the welfare of around 400 million people in about 30 years. This success was done in parallel with an increase in the number of cities within China. Between 1980 and 2011, nearly 500 million people moved to the cities creating an important source of internal demand for raw materials that began to change the dynamics of world markets. As such, this domestic demand, together with the external demand of the world markets is one of the greatest competitive advantages of China. However, at this stage of its development path, China needs to change its economic model.

Between 1949 and 1976, China has focused on modernizing its economy through large investments in the industrial sector resulting from a central planning process. This model, which has evolved over time, has allowed it to limit consumption and control savings to use the income of public companies as a source of investment to increase productive infrastructure. In turn, China focused on developing its internal potential and limiting its commercial participation in the international area. But this model changed.

In 1978, free market reforms were introduced to increase private participation in China's economy and to attract foreign direct investment (FDI). This was done under a process of gradualism where different policies were experienced within special economic zones. These policies were fruitful in that China was able to use its industrial base, its surplus labor and its economies of scale to become the world's factory: China is the main exporter and second largest importer worldwide. Although this improved the welfare of its citizens, consumption and savings remained limited. Later, China added the export component to its GDP equation. The boom in global trade since 2000 after China's entry into the WTO and the security of public investment in China by the government and the foreign private sector ensured relative stability in the development process.

China´s economy prospered until 2008. In that year, the global crisis impacts the economies of the world and especially the domestic demand of countries and important blocs such as the United States and the European Union. This generates a volatility in China, which although not affected by the crisis like the other powers, reveals an inherent fragility of its model. The fall in global demand affected the growth of its economy and the low external demand left the country with excess capacity in several industrial areas. In this way, if China cannot export, its economy slows down and the general welfare gained with so much effort is lost.

From these premises, China decides to change its economic model from growth based on investment and exports to one based on consumption and services. This change has been recognized by the government in several five-year plans with the expectation of adding a dimension of sustainability to its economic model. For China, to depend only on world exports and its internal investment is not sustainable and that is why the government is changing its traditional policies.

This structural change towards an increase in consumption implies a lower rate of growth and, consequently, an impact on external markets since China's domestic demand represents almost 50% of the global demand for many raw materials such as iron. At the same time, but it also needs other countries to grow in order to sell their products and obtain raw materials. It is in this way that the initiative of the Strip and the Route (B&R) becomes a key factor in the long-term plan of China.

Since its first appearance in September 2013, the B&R had been discussed above all in terms of promoting trade, economic development and economic growth through the interconnection of the Eurasian continent using the old silk routes. This was driven by several economic factors, such as a decline in global economic growth and trade, but more importantly, this was driven as a way for China to find the sustainability of its economic model as it moves forward with its reform.

These objectives would first be done by connecting cities by land and then doing the same by sea. From this point of view, Europe is considered the final destination of manufactured products from Asia and China as the final destination of Central Asia's natural resources. Later, it was logical to connect the maritime part to the production centers in the African continent where China already has operating facilities built through FDI flows in the continent. In this way China can take advantage of the additional synergies of its companies in Africa that incorporate these lands on the maritime road. However, the initiative has mainly focused its thinking on the Europe-China axis. However, the B&R became more than just a trade-focused initiative.

The need to expand markets in both directions to ensure sustainability provided the perfect scenario to use the initiative as a tool for the betterment of the world. Despite technological advances and the integration of supply chains, the world continues to think in terms of the cold war. In the twenty first century the world needs initiatives that put behind the individual interests of countries and begin to think about common interests: climate change and terrorism are global not local issues that can only be address as a collective. But today, instead of having more dialog between countries, the world seems to closing in. One example of this is the sudden rise in protectionism. In view of these emerging trends, the B&R initiative will help to expand connectivity between countries and offer a window to increase the dialog. This means that the initiative should not be seen only in the commercial part, but also in other dimensions that include diplomatic, economic, financial and social. As such, the B&R platform can be seen as an attempt to help to integrate the world and to improve opportunities for connectivity and peace

In this sense, it could be said that the B&R initiative is an inclusive model of development based on a win-win interest that fosters relations between nations. This is the reason why the initiative also has a cultural part that improves social connectivity among nations. Although it is difficult to separate the economic and social aspects of geopolitics, in the end, the main assumption is that the expanded dialogue among countries that are aligned under common interests, will help reduce wars and anxieties in the world.

The establishment of diplomatic relations between Panama and the People's Republic of China has changed the paradigm and broadens the potential to affect trade and other relationships throughout the world.

Panama has had relations with China since the beginning of the 19th century. There were many Chinese migrants from the southern provinces of China and the United States who came to Panama to work, first in the construction of the intercontinental railroad that connects the Pacific with the Atlantic oceans and later, for the construction of the Canal. However, relations with China were mainly commercial and limited to the extent that the US provided it, since the Canal was administered by the USA. Furthermore, China was seen as an ideological rival in the geopolitics of the Cold War. Without formal diplomatic relations with China and with its strong US presence. In turn, relations between China and Panama continued to be mainly commercial. But during that time, China was not the manufacturing power of the world.

For Panama, the establishment of relations with China comes just in time. This action offers China greater legal security for its investment and allows Panama to become an integral part of the B&R. The Panama Canal plays an important role in maritime trade between Asia and the United States and the world. It is a connectivity node that can serve as a starting point for a regional distribution center for Central and South America. This position of greater integration among the Central American markets was mentioned at the OAS meeting in June 2017.

Despite their cultural differences, China and Panama have experienced a similarity in several experiences especially with respect to the colonial enclaves that can help strengthen the relationship between countries. Between 1840 and 1949, China was at the mercy of the imperialist powers that sought extractive policies without caring about the development of the country. This began with a colonialist imposition on the Chinese ports that later expanded to various colonial enclaves since no power was able to conquer it during that period. This caused a great duality in the economy of China where an important industrial sector was generated within the area of the ports and its surroundings but it kept China in a social backwardness. Panama experienced the same thing since 1903. The country, limited by a North American colonial enclave, presented an economic duality between the country's banks and the rest of the country. Any economist who visited the areas of Clayton, Howard and Cristóbal could be confused thinking that those benefits also happened for Panama. But this was not the case.

Panama experienced years of delay and important limitations in its trade and its knowledge of the maritime area that kept the country in the group of countries in underdevelopment. Thus, the B&R initiative offers another alternative to improve the national economy and to add sustainability.

Panama is a logical extension of the silk route. The Panama Canal provides connectivity to the world through the connection of 144 maritime routes that connect 1,700 ports in 160 countries. The channel not only connects the Pacific and Atlantic markets, but also provides access to the markets of North and South America by sea, but also has terrestrial connectivity to the north and possibly connectivity with the south, since the connections with the South by land are impeded in the Darien region, but it can be developed.

One way to understand the impact of the initiative is to see China's actual water route in the container segment. On this route, the ships with manufactured products go from China to the east coast of the United States. On this route, the boats are almost full to the destination but have a problem on the return trip. Most of the boats on the return trip operate at low capacities that affect their profitability. This is a reflection of changes in the nature of goods traded between the US and China.

However, the picture changes with the B&R. As Europe becomes another distribution node in global supply chains rather than a final destination for products, the enigma of the return trip from the East Coast changes significantly. The availability of lower-priced manufactured products from Central Asia and Africa can provide additional income to increase travel profitability and strengthen North American trade opportunities to Europe. In this way, the new initiative will promote another era for the relocation of the manufacturing and integral distribution of products between countries.

The good news is that the establishment of relations with China occurs at the same time when the country approves the master plan for the development of Canal banks. This allows China to assume a more important role in its development. China has significant investments in local ports and this would facilitate maritime integration to the development of the area. Once again, it is important not to rush and develop state strategies that are in line with a long-term national vision in order to maintain regional harmony and maximize the benefits that this new opportunity offers us.

However, the B&R requires solid institutions and State policies. While the initiative is powerful, the environment of corruption and populism today presents an important constraint. This can be seen in two current examples, that of the port of Piraeus in Greece and that of Colombo in Siri Lanka, which faced the B&R in two ways. In Piraeus, the Greek government, looking after the interests of its citizens, is able to negotiate with China and obtain common benefit agreements without sacrificing its sovereignty and putting its national budget at risk. However, the government of Siri Lanka, which used Sino-Chinese relations without a strategic approach, is currently facing budget problems.

Panama has been interested in a fast railroad to the city of David. Although this seems to be an opportunity for cooperation between Panama and China, it is important that the corresponding studies be carried out. This is because the country must have clear national interests and the costs and benefits of any investment, not only from China but from any other country. If Panama seeks a long-term relationship, it should evaluate the economic and social impacts of several of the agreements it has made. Panama established a free trade agreement with the United States that has fallen short of the expectations raised about it. In addition, the experience in other countries when there is a lot of foreign investment without proper management has been the displacement of important industries at the national level and an increase in prices due to the increase in demand that negatively impacts citizens. All this points to the need for a strategic management of the relationship between China and Panama.

For this reason, it is important for Panama to strengthen institutions, reduce corruption, ensure the certainty of the punishment of the law and, most importantly, establish a strategic political group that can support decisions with China for any president. Otherwise, instead of improving conditions in the long term, the country would put at risk the advances that have been achieved so far.

The B&R Initiative provides a new platform for global integration. Not only does the framework provides support for infrastructure financing, but also the proper incentives in trade for countries to join the initiative. It needs to include the United States as well as other countries that seem to be focusing on protectionism rather than on integration. Nevertheless trade is such one dimension of the initiative. The need to reduce trade barriers will encourage negotiations and compromises that will further the dialog. These actions will help expand diplomatic, cultural and economic cooperation that can help improve the relationships among the countries of the world in the twenty first century.

34th PIANC WORLD CONGRESS PANAMA 2018 May 7 to 11

CANAL DE PANAMÁ

MARITIME SINGLE WINDOW

Republic of Panama

PANAMA CANAL SINGLE WINDOW

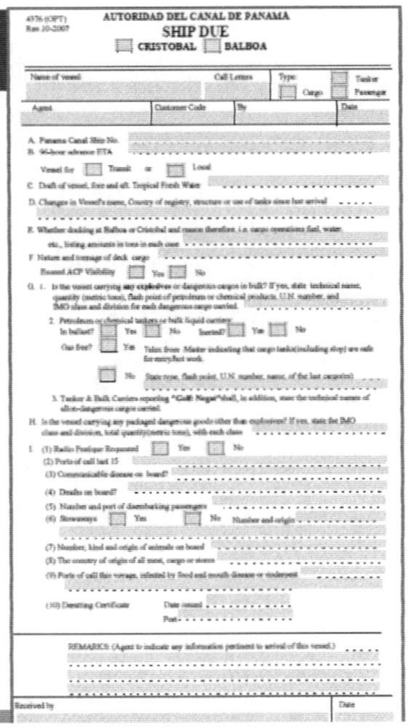
AUTORIDAD DEL CANAL DE PANAMA

SHIP DUE

CRISTOBAL BALBOA

Name of vessel | Call Letters | Type: Cargo / Tanker / Passenger

Agent | Customer Code | By | Date

A. Panama Canal Ship No.

B. 96-hour advance ETA

Vessel for Transit or Local

C. Draft of vessel, fore and aft. Tropical Fresh Water

D. Changes in Vessel's name, Country of registry, structure or use of tanks since last arrival

F. Nature and tonnage of deck cargo

REMARKS

Received by | Date

Panama Canal Single Window - WEB

Electronic Data Collection System

2004 – 2015

Oracle	Java

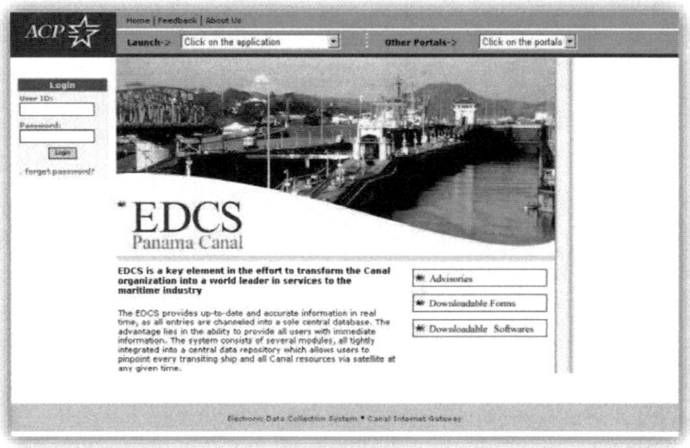

- ❑ WEB
- ❑ Offline
- ❑ B2G Gateway

Panama Canal Service Portal

2015 -

Oracle Fussion Middleware

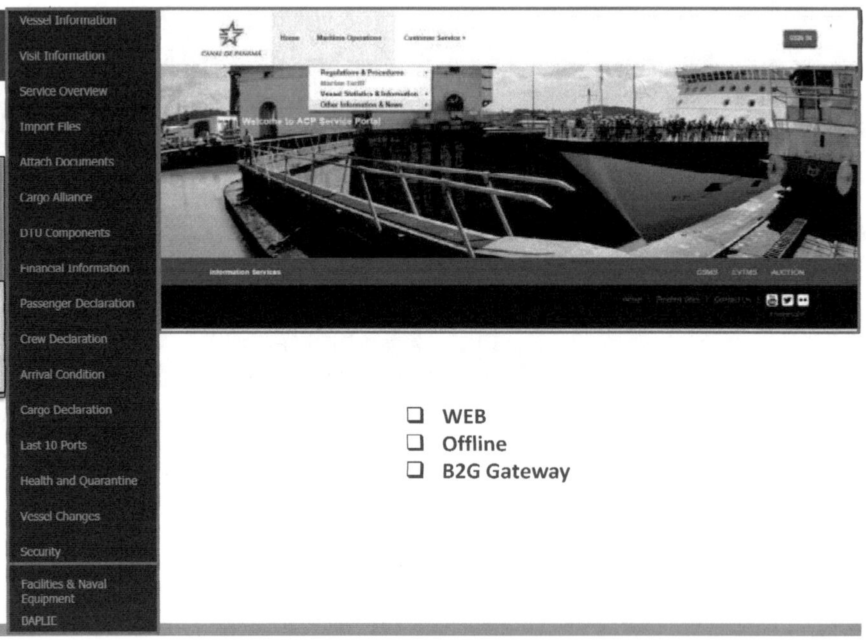

- ❑ **WEB**
- ❑ **Offline**
- ❑ **B2G Gateway**

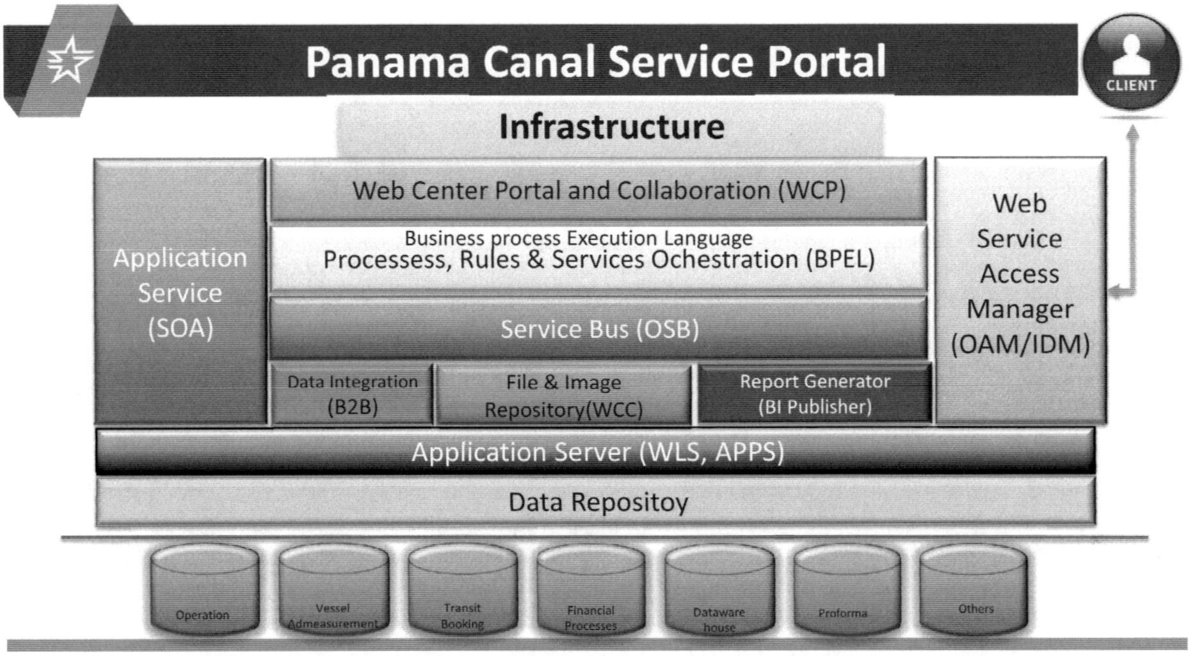

Panama Canal Service Portal
CLIENT
Infrastructure
Application Service (SOA)
Web Center Portal and Collaboration (WCP)
Business process Execution Language
Processess, Rules & Services Ochestration (BPEL)
Service Bus (OSB)
Data Integration (B2B)
File & Image Repository(WCC)
Report Generator (BI Publisher)
Web Service Access Manager (OAM/IDM)
Application Server (WLS, APPS)
Data Repositoy
Operation
Vessel Admeasurement
Transit Booking
Financial Processes
Dataware house
Proforma
Others

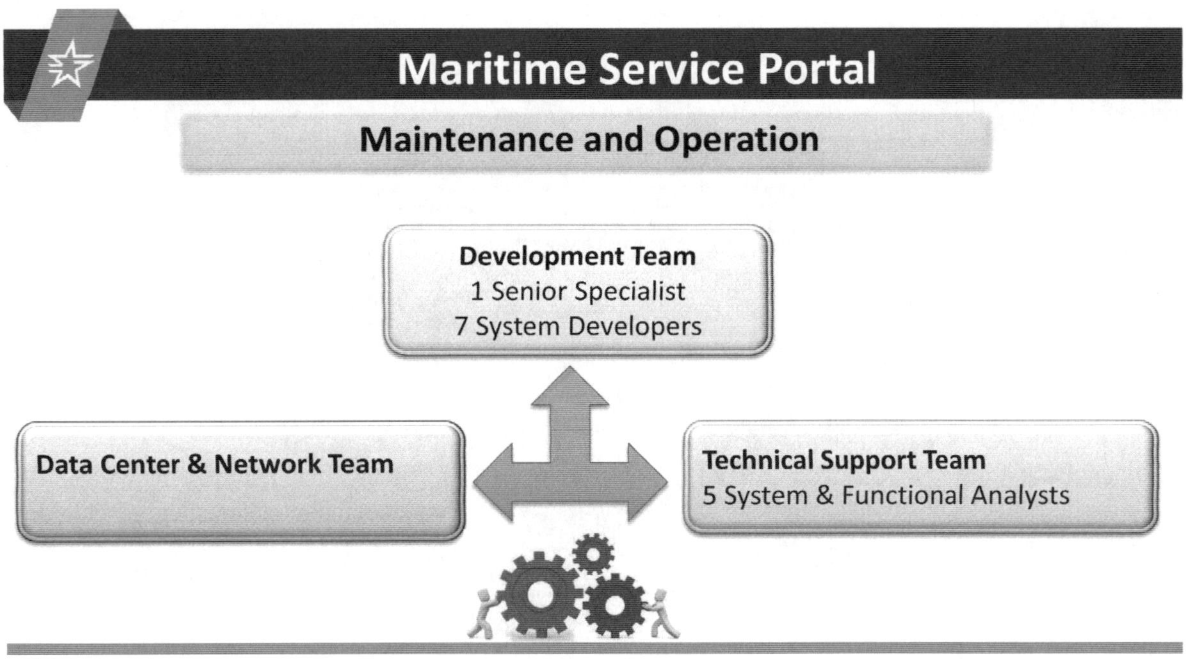

Maritime Service Portal
Maintenance and Operation
Development Team
1 Senior Specialist
7 System Developers
Data Center & Network Team
Technical Support Team
5 System & Functional Analysts

Historic References

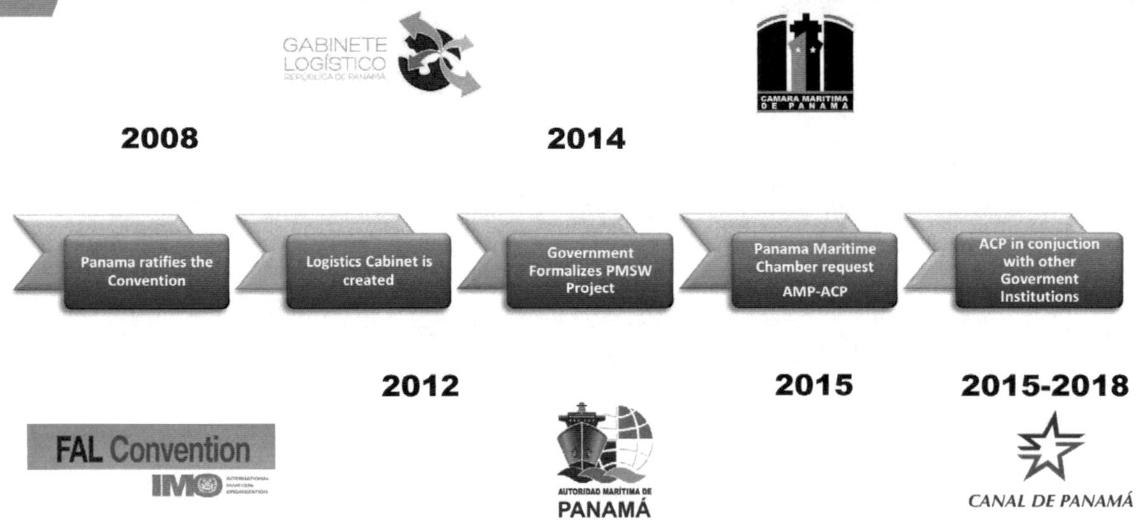

Improvement Opportunities

Non-standard complex processes for arrival and departure formalities

Public and private entities do not share information

Delays in Loading, unloading and transshipment operations

Paperwork via physical collection and manual processes

Implementation Issues

Technology

- Distrust in the use of technologies
- Low investment level
- Different level of technology developed in each government agency

Legal

- Divergent regulations
- Old procedures included in regulations
- Conflicts due to overlap of responsibilities

Resistance to Change

4 Strategy Implementation Axes

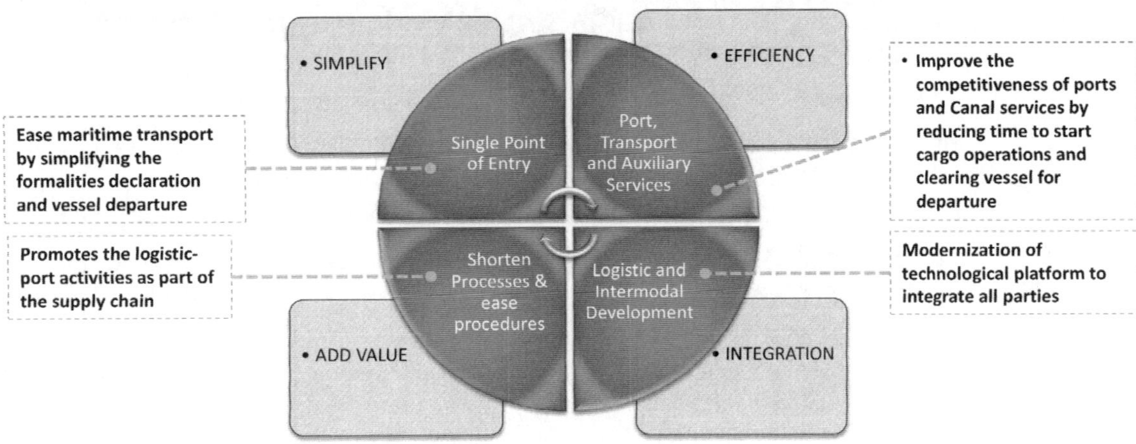

Integrated Maritime Community System

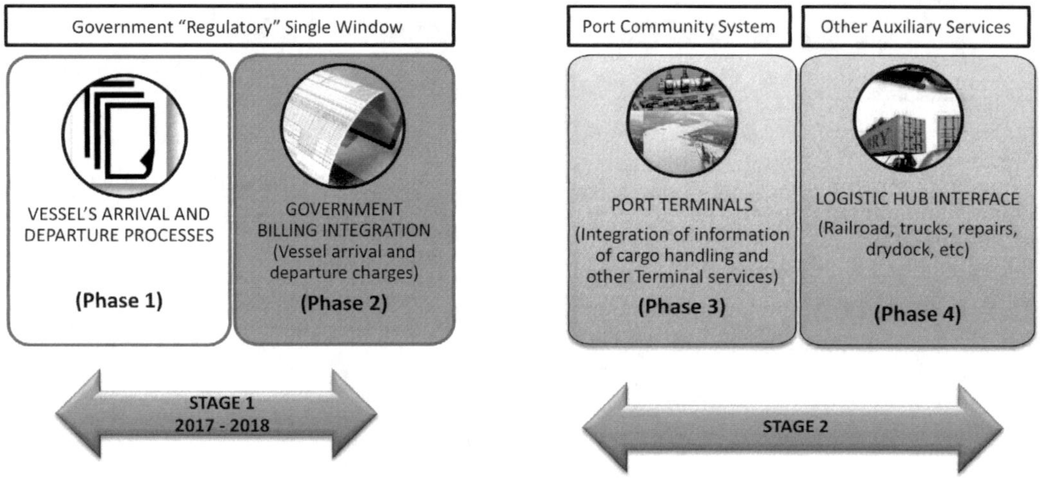

GAME CHANGERS

- ✓ Use the Panama Canal Single Window as the PANAMA MARITIME SINGLE WINDOW "VUMPA"
- ✓ Panama Canal Authority operates and maintain the Single Window
- ✓ One Single Vessel´s Arrival Inspector and share its inspection result
- ✓ Risk Assessments prior to Vessel´s Arrival

MARITIME SINGLE WINDOW OF PANAMA

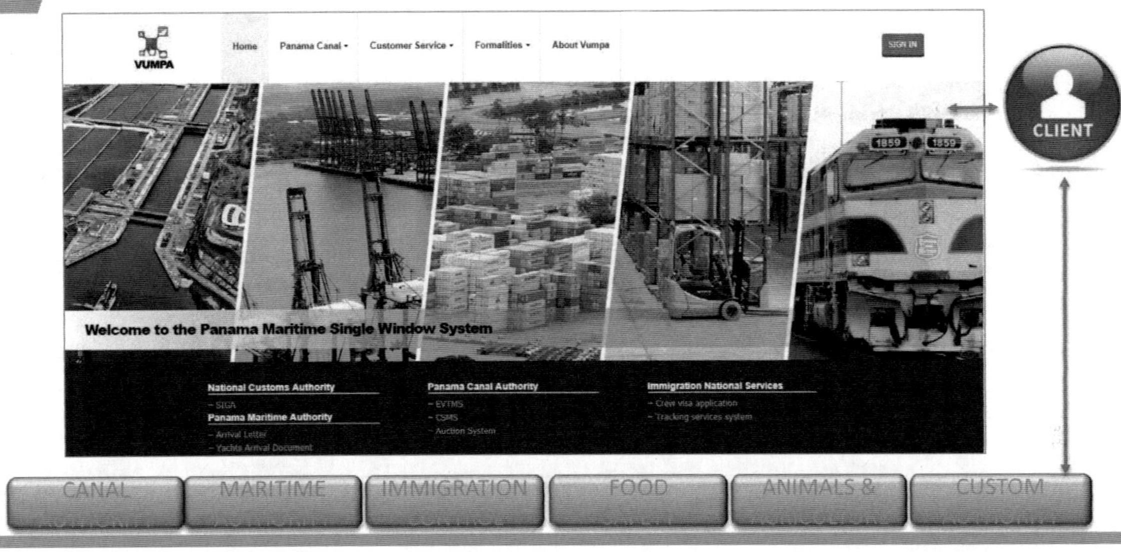

Government Regulatory Compliance

Visit No.	Vessel Name	ETA	Arrival Location	Begin Location (1st Mov.)	End Location (1st Mov.)	MINSA		AMP		MIDA		ANA		SNM		PANAMA
						PRE	ARRIVED	PRE	ARRIVED	PRE	ARRIVED	PRE	ARRIVED	PRE	ARRIVED	Inspection Completion Date
323132	SUNLEAF G...	26-Apr-2018 ...	Pacific	PSEA	ASEA	CLEAF	SELEC	SELEC	SELEC	SELEC	SELEC	SELEC	SELEC	SELEC	SELEC	
323980	GREAT PRO...	26-Apr-2018 ...	Pacific	PSEA	AAN	SELEC	SELEC	CLEAF	INPRC	SELEC	SELEC	SELEC	SELEC	CLEAF	SELEC	
324197	ORANJE	26-Apr-2018 ...	Pacific	PSEA	PDK	CLEAF	SELEC	CLEAF	SELEC	SELEC	SELEC	SELEC	SELEC	CLEAF	SELEC	
323128	G. SPIRIT	26-Apr-2018 ...	Pacific	PSEA	ASEA	CLEAF	SELEC	SELEC	SELEC	SELEC	SELEC	SELEC	SELEC	SELEC	SELEC	
309959	MSC CATER...	26-Apr-2018 ...	Pacific			SELEC	SELEC	SELEC	SELEC	SELEC	SELEC	SELEC	SELEC	SELEC	SELEC	
324014	APPEL DAIR	26-Apr-2018 ...	Atlantic	ASEA	SHBMAR	SELEC	SELEC	SELEC	SELEC	SELEC	SELEC	SELEC	SELEC	SELEC	SELEC	

Page 1 of 7 <<First Previous 1 2 3 4 5 6 7 Next Last>>

BENEFITS

VUMPA
Panama Maritime Single Window System

- Reduces customer costs through the use of an electronic declaration process
- 6,000 hours saved at the port terminals
- Risk assessment performed prior to vessel´s arrival
- Single point of entry to declare government formalities
- Reduces around 300,000 forms, documents and paperwork per year
- Pre-arrival, arrival and inspection information is shared between the Government entities
- Automatic Free Pratique allows bunkering services prior to inspection

Lessons Learned

Support from top government officials

Formalize a port logistics community

Ensure the regulations for the Project

Support of the private sector

Involve stakeholders in the initial phases of the Project

Communication, Change management and follow up
"individual response to change"

NEXT TO COME

- Single Departure Process – Electronic Form
- Additional Cargo declaration formats
- Share information with Port Terminals
- Simplify the vessel´s fumigation process
- Communication with other Maritime Auxiliary Services

Automate the Government Maritime Service Charges "Arrival and Departure Services"

PANAMA MARITIME SINGLE WINDOW

CANAL DE PANAMÁ

SHIPPING LNG FROM AN ARCTIC LNG PLANT: SOME MARINE CHALLENGES

by

Frederic J.L. Hannon[1],

ABSTRACT

Novatek, Total, CNPC and Silk Road Fund are partners in the Project named Yamal LNG located on the Yamal Peninsula (Russia) to produce the huge gas reserves of South Tambey fields (about 927 BCM of natural gas). The LNG Plant is located in Sabetta on the west bank of the Ob River, North of the Arctic Circle. The extraction of gas and the building of a liquefaction plant on permafrost in these regions seem to be less challenging than its transportation under LNG form, from the production plant to the markets, as the LNG & gas condensate carriers have to use the Northern Sea Route, in ice-covered waters, for about nine months a year. A year-round accessible and operable Port therefore had to be built.

To evacuate the 16,5 Mtpa LNG and 1,2 Mtpa of gas condensate production from the Liquefaction Plant, the shipping solution chosen to ensure the safe and reliable maritime transportation year-round was to build a fleet of up to 15 Arctic LNG carriers, with sufficient ice-class notation (Arc7) and ice-breaking capability to operate without the assistance of the Russian ice-breakers in the conditions of Barents and Kara seas. They will operate independently year-round to North-West Europe where the cargoes will be sold or transferred unto conventional LNG carriers at selected trans-shipment terminals.

To accommodate these vessels in Sabetta, the new Port had to be designed in such a way that it is accessible and operable year round. This means that compared to conventional marine facilities, the focus has been put on the ice management of the navigational waterways – dredged river channels and Port access channels, the sheltering of the jetties of the Port by ice protection barriers, the jetties and quays designed to sustain the ice loads, their winterization and their ice management through brash ice management system, and the sizing of a support fleet in order to ease the operations in the ice.

The lessons learnt during the construction phases of the Plant – dredging of channels, delivery of materials for site preparation and civil works during open water season and in the winter, building the marine offloading facilities for delivery of the modules of the Plant, steps to the final design of merchant ice-breaking ships, construction of a future international airfield, current and future logistics solutions for using the Northern Sea Route for the Project purpose – and much more challenges are described in the paper.

The first Ice-breaking LNG Carrier was delivered in 2016 and has carried-out her ice-trials successfully in March 2017 ; the first liquefaction train of the Yamal LNG Project started in December 2017 and in March 2018, the first million tons of LNG was safely delivered to the gas buyers.

---

[1] TOTAL S.A., Gas, Renewables and Power, LNG Shipping, France, frederic.hannon@total.com

1. INTRODUCTION

The Yamal LNG Project is definitely a pilot project for Russia: initiated by a decree from the Russian government on October 11th 2010, which promoted the integrated development of the LNG Industry, it also granted some exemption taxes on producing and exporting LNG and condensates, participating in the building of new infrastructures such as the marine Port and its access, an airport, and providing all necessary framework to expedite the materialization of the Project.

The final investment decision was taken in December 2013 by the sponsors of the Project and after four years of construction, the first train of production started in December 2017. The two other trains will successively be added with a one-year timelap to reach the full production of 16,5 Mtpa of LNG and 1,2 Mtpa of gas condensates. The milestone of the first million tons of LNG produced was announced on March 2018; today Sabetta Port is an international maritime port, with an international airport, in the North of the Arctic Circle (Fig.1).

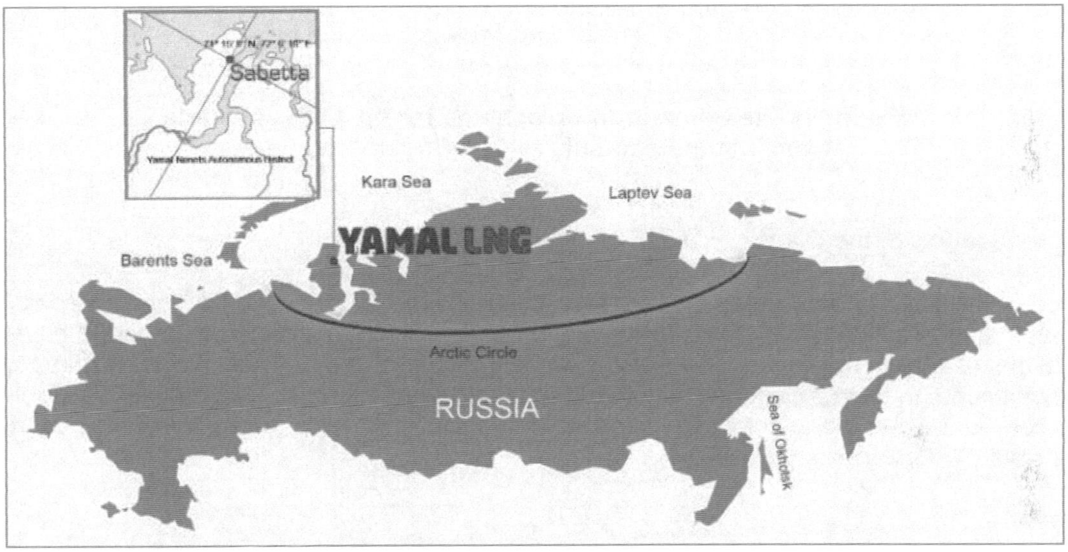

Figure 1: The location of the Yamal LNG Project, North of the Arctic Circle
Courtesy of Yamal spg

2. AN INTEGRATED PROJECT

2.1 The general scheme for shipping the LNG and gas condensates

Due to the location of the Liquefaction Plant on the Yamal peninsula, the shipping solution for reaching European and Asian markets on a safe, regular and sustainable basis is to consider using the Northern Sea Route, to the West all year-round and to the East during the open-water season (i.e. five months per year approximately).

A dedicated fleet of Arctic LNG carriers with sufficient ice-breaking capabilities to overcome the annual tear ice of the Kara and Barents seas and sufficient ice-class notation to be authorized to navigate independently by the Northern Sea Route Administration had to be built. After numerous simulations and taking the assumptions from the marketing, 15 Arc7 LNG Carriers of about 170,000m^3 of cargo capacity have been ordered to deliver the total production of the three trains, that is 5,5 million tons each per year.

For the deliveries to Asian gas buyers during the ice congested period of the eastern seas of the North East Passage, a trans-shipment of the cargo between Arc7 LNG carriers and conventional LNG carriers has been considered in Europe (Fig.2).

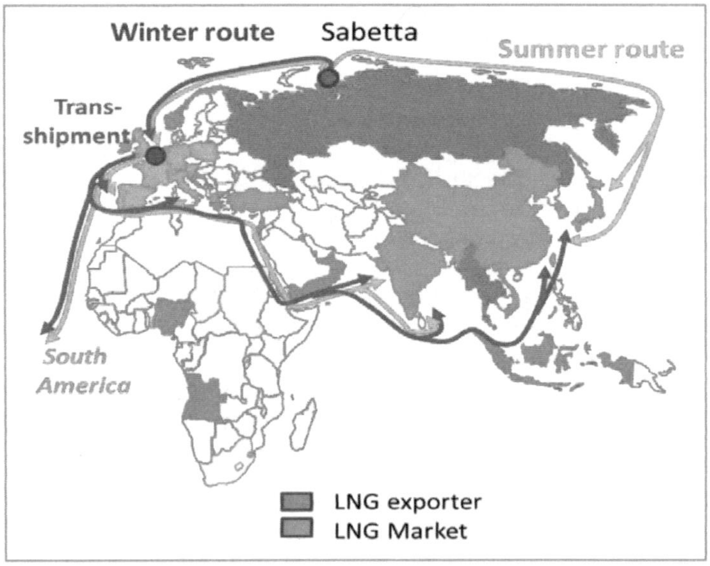

Figure 2: The transportation scheme for the LNG products from Yamal Peninsula to Gas Buyers' markets

2.2 The location of the Plant

The choice of the plant location has been dictated principally by the situation of the gas fields of South Tambey (gas reserves 2P PRMS: 926 BCM on 31.12.2013) : the Plant and the Port had to be as near as possible to the reserves to avoid any long pipeline feeding the Plant. A small port in Sabetta already existed, created by pioneers in the 1980s to accommodate geological survey teams. In the year 1985, some trials had been carried out by experimental expeditions to prove the feasibility of navigation in the ice of the Kara sea in the area (ARCDEV Program) (Fig.3).

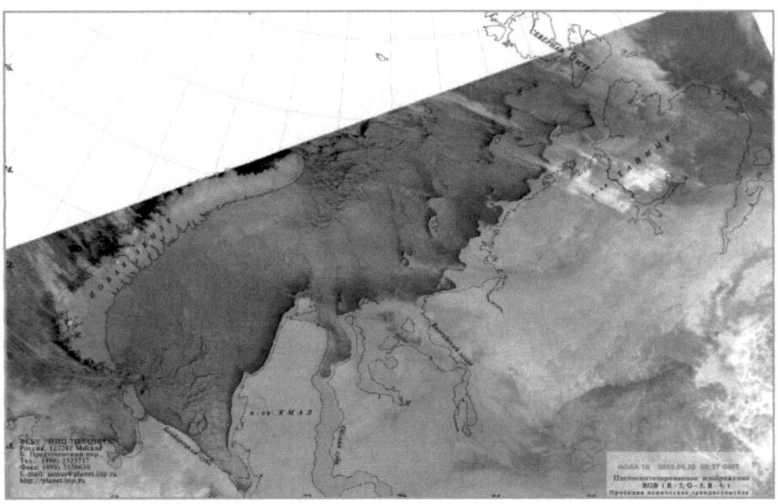

Figure 3: The Yamal Peninsula with the Ob River mouth in fast ice (February 2013)
Courtesy of AARI

During the first years of site preparation, due to the remote location and the absence of suitable year-round overland transportation infrastructure, most of the raw materials and equipment required were delivered to site by sea, using the Northern Sea Route. The old way of unloading equipment and materials on the ice and building roads (Fig.4) was used in order to start building the Marine Offloading Facilities (MOF) of the

Port; it also allowed to collect additional data on local ice conditions during the course of the vessels' voyages and during the year in the area of the future Port .

Three different ice periods have been identified off Sabetta: a "freeze-up" period during which ice first forms and thickens but is still mobile (starting beginning of October); the "fast-ice" period during which the ice is not moving but is still growing (about 3 weeks later); the "break-up" period with thick floes drifting in the Ob Gulf and which could invade the Port (about three weeks in July). This is representative of a typical winter. At the end of the ice period, maximum estimated ice thickness can be up to 2,5 meters at the mouth of Ob River.

Figure 4: Unloading on ice with the multipurpose vessel SA15, offshore Sabetta (courtesy TOTAL)

The MOF will allow, later on, to accommodate the heavy lift vessels and modules carriers delivering the components of the Liquefaction Plant. The main philosophy to build the Plant is a modular concept, with most of the modules assembled in East Asia; this would reduce the impact of the polar environment on the construction process, enable the best use of available labor, and take into account the remoteness and limited access to the site. The modularization allows construction works to be carried out simultaneously in several locations, with the modules being delivered and connected on site, hence minimizing hook-up, pre-commissioning and commissioning activities at the site.

3. THE CHALLENGES OF THE PORT AND OF ITS ACCESS

3.1 The North Access channel and Port channel

From the origin, the small Port of Sabetta is located on the West Bank of the Ob River (71°15'N,72°E) For large ships with a draft of about 12,5m to access the location, a channel of about 50 km had to be dredged in the North part of the river. How to design this channel for its use in open water and in ice conditions ?

In ice season, after each passage of a vessel, the ice will quickly grow back in the open track, meaning that after several passages, the ice thickness in the channel will be bigger than natural ice growth thickness, resulting in higher resistance for the ship. After a period of time, the channel becomes clogged and non-navigable; another track has to be used.

Several trials have been carried out in Lula (Sweden) in order to assess the phenomenon of brash ice formation in broken channel with an ice-breaker in order to provide data to correlate and validate some brash ice growth models (Fig.5 & 6).

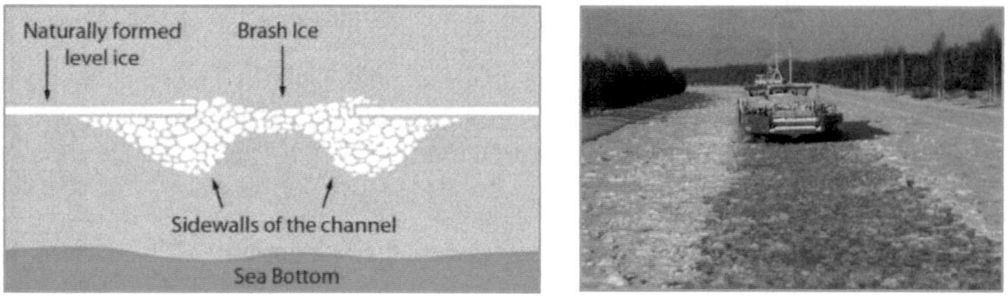

Figures 5: The formation of brash ice studied using ice breaker in a channel

The final design of the North Access Channel encompasses the need of three to four different lateral tracks in the dredged channel during the same winter (Fig.6). This also depends on the actual traffic of the vessels during the next years.

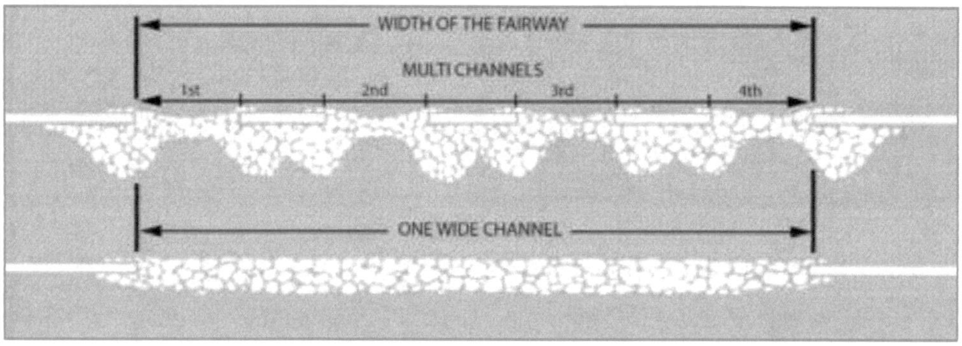

Figure 6: Multi-channels approach for the North Access channel to Sabetta

The access channel to the Port of Sabetta also needed to be dredged (Fig.7).

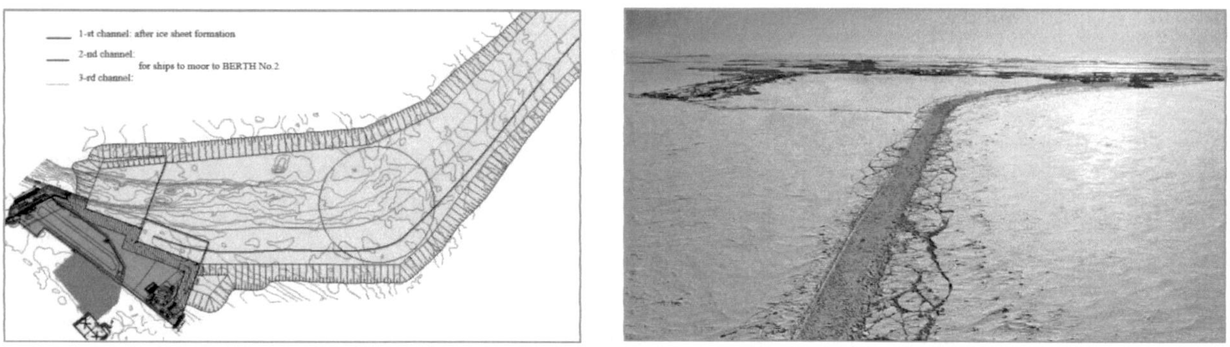

Figure 7: Access channel to the Port of Sabetta

3.2 The dredging works

An armada of dredging vessels have started the works in 2012 offshore Sabetta, in order to the construct the quays of the Marine Offloading Facilities (Fig.8).

Figure 8: Dredging works in access channel near MOF

3.3 The Sabetta Port lay-out

The review of existing ports in similar Arctic conditions (examples of Dudinka in Ienissei River and Vitino) has concluded that the export jetties of the future LNG Plant have to be protected from currents and drifting ice in order to be operational year-round.

One design requirement is to have a water depth of 15m to ensure safe access and maneuverability of large vessels in the Port. The 15m isobath is located about 7,5km from the shoreline near Sabetta. The compromise between long marine structure and limited dredging versus jetties close to shore with significant dredging , led to an optimized option of marine infrastructures located close to the shore with two ice protection barriers, one at the Northern side (about 1,850m long), the other at the Southern side (about 2,000m long). The turning basin is located inside the Port and the two export jetties are built laterally to the southern breakwater: the entrance width of the Port is about 500m and the access channel requires a length of about 5km to be dredged (Fig.9).

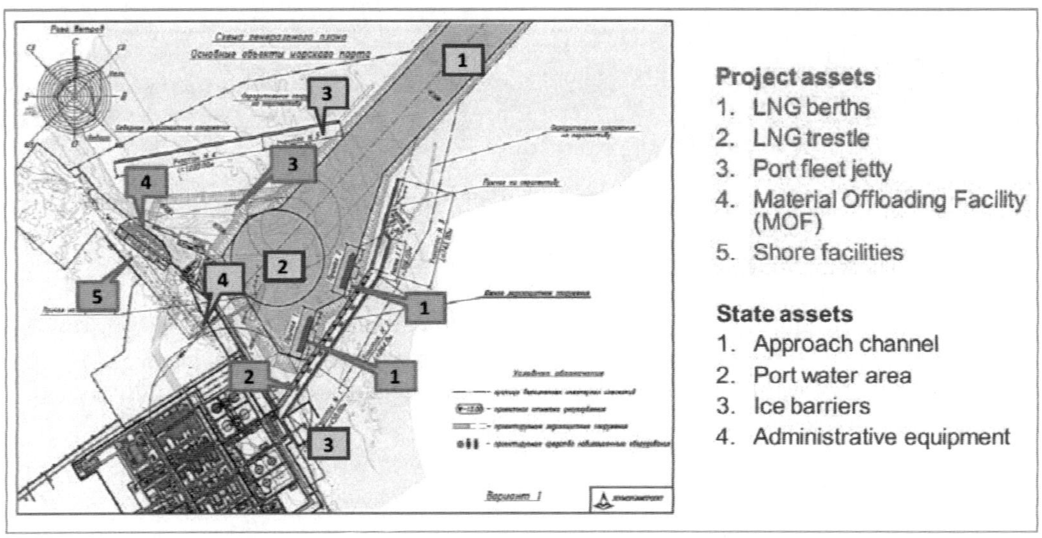

Figure 9: The general lay-out of the Arctic Port

The development of the Sabetta Port started with a first stage: the dredging works offshore of the future MOF (Fig.10) in order to build the four quays for unloading the multipurpose vessels and module carriers and accommodate the future Port support fleet (4 tugboats and one port ice-breaker).

Figure 10: North Channel (left) and Construction of Marine Offloading Facilities

In a second stage, the main jetties for exporting LNG and condensates had to be built at the same time as the storage tanks and Liquefaction Plant : Jetty 1 for LNG and Jetty 2 for both LNG and gas condensates (Fig.11). The main jetties are protected by breakwater and ice protection barriers in order to allow the loading operation of the ships all year round, unexposed to the river currents and drifting ice blocks during the break-up period.

Figure 11: Construction of the South-West Breakwater and Ice Barrier & Main Export Jetties 1 & 2

3.5 The ice management in the Sabetta Port

A lot of experience has been accumulated since the first year of multipurpose ships calling at Sabetta during winter season, in particular for the ice management. The use of tugboats and ice-breakers ahead of the arrival of the ships as in other subarctic ports has been the primary tool to make the quays of the Marine Offloading Facilities ice-free and accessible.

Figure 12: Ice management by ice-breaker during MOF construction

However, some of the corners and extremities of the quays were still clogged with ice and difficult to reach by the support vessels without the risk of hurting the quays and deteriorating the fenders (Fig.13). This could jeopardize the delivery schedule of the module carriers.

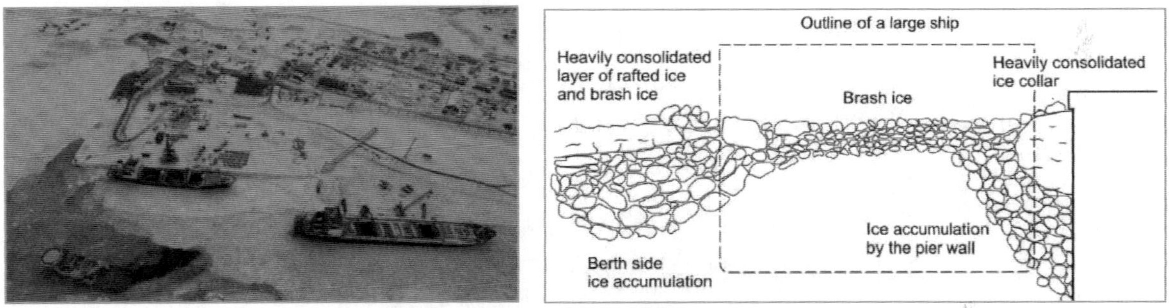

Figure 13: Ice accretion on MOF quays

An additional brash ice management equipment was tested to mitigate this issue.

A pilot mobile brash ice management system (BIMS) was installed at one quay of the MOF for assessing the efficiency of such system and size a potential global system for the area of main jetties of the Port (Fig.14).

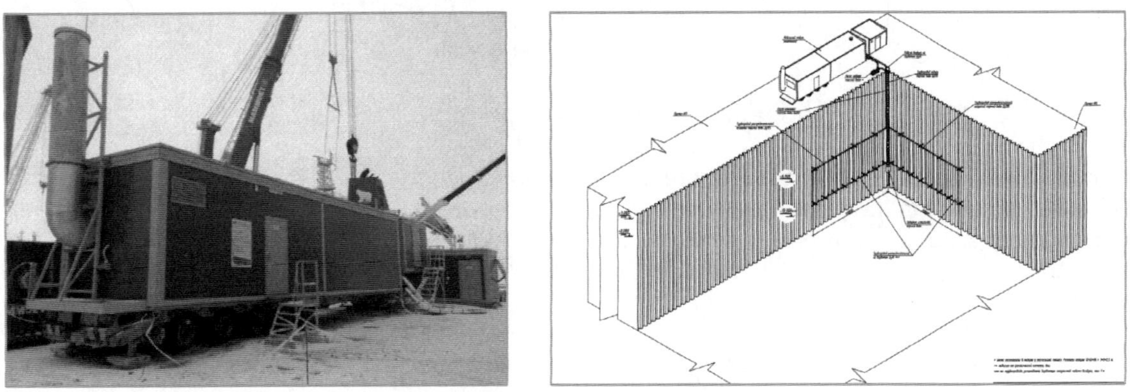

Figure 14: Pilot mobile BIMS

The project team has visited quite a lot of ice infested ports in Finland to witness the mitigation measures used to keep the accessibility of ships during winter seasons (e.g. Vuosaari). Some have adopted a brash ice management system using recovered energy from nearby power plant to heat the water of some part of the Port and slow down the growth of brash ice. A system of compressed air generating bubbles and helping to diffuse the heat is implemented (Fig.15).

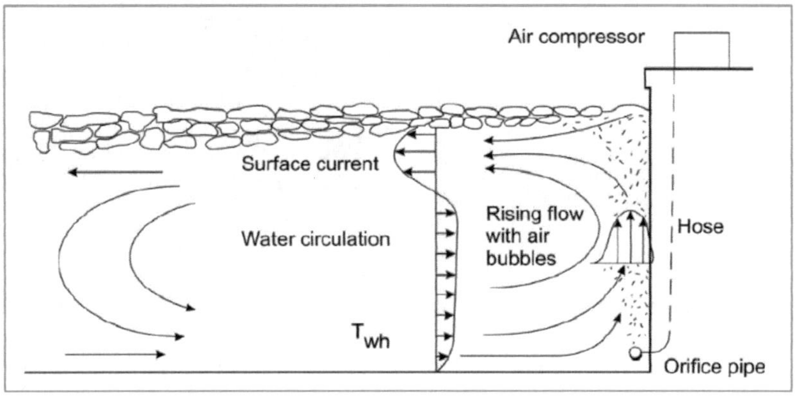

Figure 15: BIMS principle with warm water and bubbles

However, this system has not yet been implemented for a port as large as the Sabetta one.

Assuming the final layout of the Port, the local ice conditions and the anticipated maritime traffic in the Sabetta Port, a campaign of simulations has been run in order to size the equipment and assess the impact of such BIMS (Fig.16).

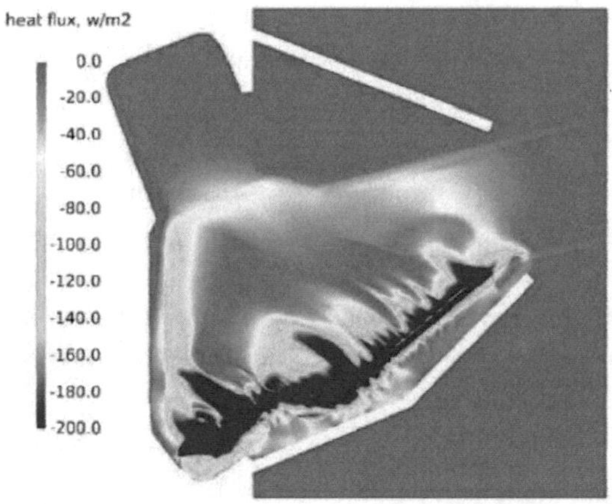

Figure 16: Simulation of heat diffusion of a BIMS inside the Port

However, as the ice management needs an integrated solution encompassing the operations of ice breakers and tugboats, as well as the actual contribution of maritime traffic and the maneuvering tactics, the heat provided by the ballast system of the vessels, and the supplement to be provided by the BIMS, the final design of the equipment and its installation has not yet been decided at the time of the first train commissioning.

3.6 The Sabetta Port support fleet

For the starting of the commercial operations of the LNG Plant, a dedicated support fleet was defined, consisting of 4 tugboats with reinforced hull and fire-fighting capabilities, able to handle the large vessels and assist the ice management, and a Port icebreaker.

	Ship	Propulsion Power, mW	Ice class
1.	Port Icebreaker"ARC - 124"	12	Icebreaker - 7
2.	Icebreacking tug "T 40105"	7	Arc - 6
3.	Escort tug "T3150A" - "Пур"	4	Arc - 4 (Hull 5)
4.	Escort tug "T3150A" - "Тамбей"	4	Arc - 4 (Hull 5)
5.	Icebreacking tug "T 3687 "	6,4	Arc - 6

Figure 17: Dedicated Support Fleet of Sabetta Port for the Project

4. THE MODULE LOGISTICS FOR THE LIQUEFACTION PLANT

One of the biggest challenges of this Arctic project has been the safe and reliable transportation and timely delivery of Plant modules; no piece of the puzzle could be missing. Indeed the modules have been built mostly in East Asia and delivered in a similar way as the future LNG export scheme.

About 152 large modules, representing about 360,000 metric tons with 55,000 tons of reinforcing and sea fastening), coming from 10 different yards in China or Philippines have been loaded on a fleet of 23 heavy lift vessels and modules carriers. Two dedicated Arc7 module carriers have been specially built in China during the first years of the project to allow an enlarged window of deliveries on site in winter conditions. A special Fleet Operations Center has been set to monitor the routing of the vessels with regard to the ice conditions, providing the data for decision and route changing should any risk be identified during a voyage: whether to follow the route through the Western route (Suez Canal) or be ready to enter the Eastern route (Bering Strait) (Fig.18). 77 shipments of large modules have been organized, with a total of 22 voyages through the Bering Strait.

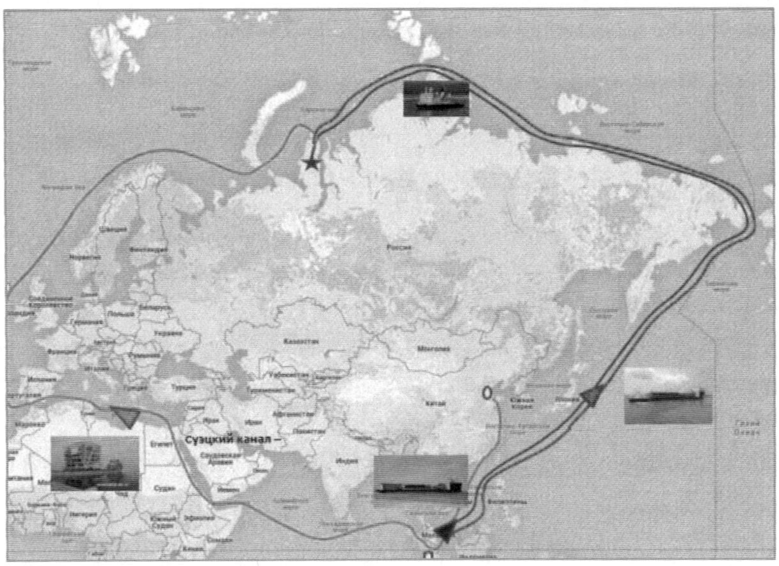

Figure 18: The module carriers fleet routing from Asian yards to delivery at Sabetta

A MISY (Module Intermediate Storage Yard) had been open in the Zeebrugge Port (Belgium) in order to serve as a buffer for the modules before final transfer to the site, especially in winter season (Fig. 19 & 20).

Figure 19: Plant Module leaving the MISY of Zeebrugge towards Sabetta

Figure 20: Module unloaded at Sabetta on a SPMT (Self-Propelled Modular Transporter)

5. THE DESIGN OF THE LNG CARRIERS AND CONDENSATE TANKERS FLEET

From the initial studies for the shipping scheme, the ships designed for the export of the production of the LNG Liquefaction Plant should be large enough for economies of scale and able to be independent from the continuous assistance of ice-breakers during their voyages along the Northern Sea Route. Their large size and width would have required two ice-breakers to open a track in the ice.

After extensive studies of existing options of vessels in service in Arctic regions and innovative solutions, the technical specifications of the first Ice-breaking LNG Carrier were finalized : a Double-Acting System type ship (Fig.21), capable of sailing both ahead and astern, having an optimised open water ice-breaking bow for both open water and light ice condition and an optimised ice-breaking stern when the ice conditions are more severe.

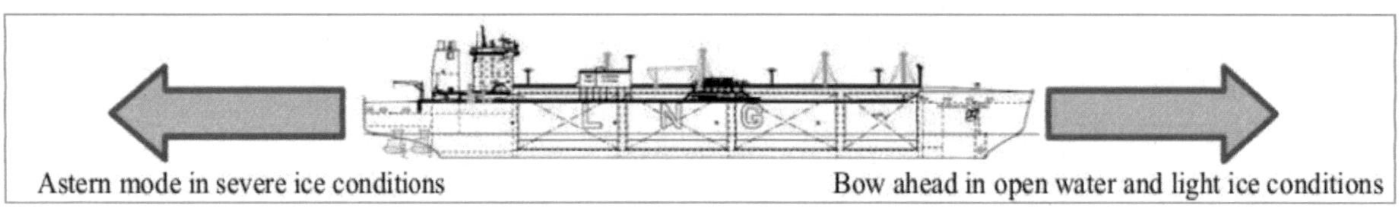

Figure 21: Double-Acting System ship

An optimized hull form and innovative and powerful propulsion plant consisting of three azimuthal thrusters of 15 MW each will allow to break the ice and maneuver in the icy waters. A reinforced hull with an Arc7 ice class notation, and double hull structure will ensure adequate safety to the cargo containment system of the ship. The boil-off gas of the LNG inside the cargo tanks will be used for the feeding of the Dual Fuel Diesel engines (6 units) for powering the hotel load, the propulsion and all cargo handling operations. The winterization of the ship will allow operations at very low ambient temperatures, down to -52°C (Fig.22).

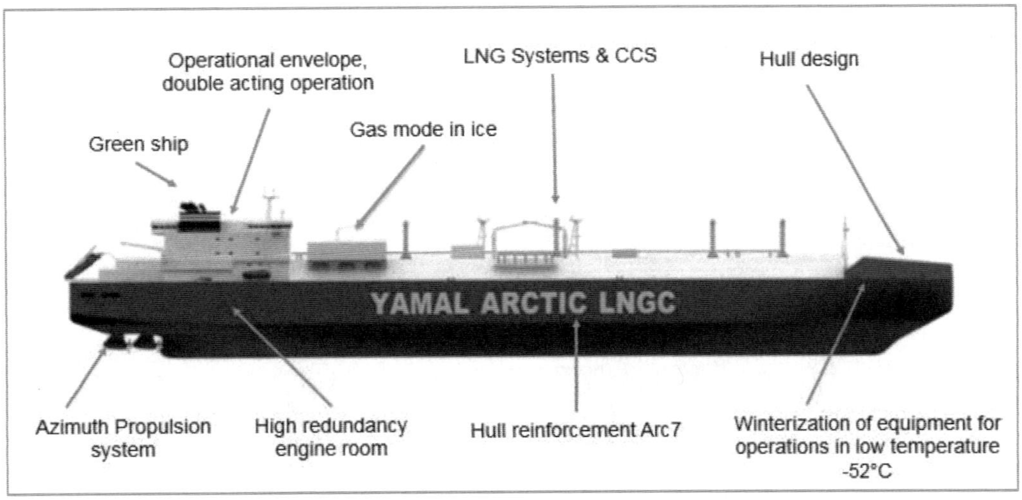

Figure 22: The Arc7 ice-breaking LNG carrier

Fifteen of such Arc7 LNG Carriers have been ordered to a Korean shipyard (DSME) and the first ones have already been delivered at the end of year 2017 and deployed to service the production of the first Train of Yamal LNG. The first one, named "*SCF Christophe de Margerie*" (Fig. 23) has successfully carried out a full campaign of ice trials in the Arctic, in March-April 2017, demonstrating higher performances and ice-capabilities than expected: the ship is capable of sailing at more than the expected figures of 2,0 knots ahead and more than 5,0 knots astern in 1.5 m thick level ice, and has better maneuvering capabilities than estimated.

Figure 23: The SCF Christophe de Margerie, first Ice-breaking LNG carrier of the Yamal trade Fleet
Courtesy of DSME

With the same assumption for the design, two Arc7 Condensate Tankers are being built for the project, one in Finland, one in China to start the export in 2018.

6. THE TRANS-SHIPMENT TERMINALS IN EUROPE

As part of the transportation scheme, some potential trans-shipment terminals in Europe have been assessed during the phase of construction of the Arc7 LNG Carriers. The project has a long term agreement with the Terminal of Zeebrugge, where an additional storage tank has been erected and some dedicated equipment fitted. The trans-shipment can be from ship to ship through the jetty or from the ship to storage to ship.

Other alternate trans-shipment terminals can be also used, such as Dunkirk LNG (Fig.24), Montoir de Bretagne (France) (Fig. 25), or Gate LNG (Netherlands), as the LNG Carriers are compatible with them.

Figure 24: Trans-shipment at Dunkirk LNG
Courtesy of Dunkirk LNG

Figure 25: Trans-shipment at Montoir de Bretagne between an Ice-breaking LNG Carrier and a conventional LNG Carrier
Courtesy of Nantes Saint Nazaire port

7. CONCLUSIONS

Accordingly, a lot of challenges have been addressed in order to launch the Yamal LNG Project.

All the lessons learnt during the various phases of design, construction, and commissioning of the Plant will provide solid foundations for future projects and serve as a reference for others who want to develop and exploit the large gas reserves of the Arctic area.

Figure 26: View of the Yamal Project in Sabetta

-

References

Hannon, F., Kalinin, A. & Giboin P. (2015). Navigating the NSR: The Arctic Ice-Breaking LNG Carrier. OTC 25478

Coche E. & Kalinin.A. (2013). Yamal LNG: Challenges of an LNG Port in Arctic – POAC'13 – Espoo, Finland

Hannon, F. & Bogdanovskiy K. (2016). Navigating the Arctic. *LNG Industry Magazine.* July 2016.

PIANC (2004). INCOM Report WG 23,Technical and Economic Problem of Channel Icing, PIANC, Brussels.

PIANC (2014). MARCOM Report WG 121, Harbour Approach Channels Design, PIANC, Brussels.

PIANC (2018). MARCOM Report WG 185, Port on Greenfield Sites, PIANC, Brussels.

http://yamallng.ru/en/

TRANSPORTATION INFRASTRUCTURE AND CARGO LOGISTICS MASTER PLAN FOR THE INTEROCEANIC ZONE OF THE PANAMA CANAL

by

Rebeca Caceres[1] *and Miguel Arosemena*[2]

Keywords: master plan, economic zones, value added logistics, maritime hubs

ABSTRACT

Approximately, three percent of international seaborne trade flows through the Isthmus of Panama. Likewise, fifty per cent of Latin American transshipment and eight per cent of import flows to the region, currently take place within the Interoceanic Zone of the Panama Canal (PMZIC). Less than one percent of these international flows are transiting or receiving services within the territory of Panama. However, its contribution to the Panamanian economy is estimated at one third of its total value.

With the aim to maximize these benefits, the Panama Canal Authority and the Government of Panama carried out the Transportation Infrastructure and Cargo Logistics Master Plan for the Interoceanic Zone of the Panama Canal. This Master Plan is the first integrated transportation infrastructure and cargo logistics planning effort in the country. The study identifies the main challenges for capitalizing on surrounding areas of the Panama Canal (around 75,000 hectares), and proposes a set of action in infrastructure, services and implementation aspects to offer an efficient platform for value added logistics to contribute to the Canal´s and conglomerate land profitability.

1. INTRODUCTION

The Transportation Infrastructure and Cargo Logistics Master Plan for the Interoceanic Zone of the Panama Canal is the first integrated transportation infrastructure and cargo logistics planning effort in the country. The need for this master plan was identified in 2015 ´by the Logistics Cabinet´s Infrastructure Committee, which is a part of the Logistics and Competitiveness Secretariat of the Ministry of the Presidency.

This committee is currently headed by the Panama Canal Authority and is composed of public institutions and logistics representatives from the private sector. The document was developed jointly with the National Logistics Strategy, led by the Logistics Cabinet.

The objective of this Master Plan is to generate long-term, sustainable conditions for the development of Panama as a global logistics hub. At the same time, to determine priorities for investment in freight infrastructure in order to provide better integration, optimizing the existing infrastructure and generating opportunities to capture new cargo segments with appropriate value-added logistic services, and at the same time to ensure the preservation of environmental resources and public and urban spaces within the Interoceanic Zone of the Panama Canal.

Through interinstitutional consultation, relevant databases were obtained from various national agencies, as well as relevant information from project developers and stakeholders within the zone as

[1]Liner Segment, Panama Canal Authority, rebcaceres@pancanal.com
[2] Business Development Specialist, Panama Canal Authority, marosemena@pancanal.com

UABR[3], Free Trade Zones and Special Economic Areas as Panama Pacífico Agency and Colon Free Zone, Tocumen S.A., the Municipality of Panama, Miviot[4], Mici[5], Mef[6], among others. During the formulation, the study was continuously validated through four workshops, a focus group, direct and indirect interviews, and field surveys with stakeholders.

The first stage of the study consisted of an assessment of the current condition of the hub, in order to complete a gap analysis. For this, information was gathered from physical and digital inventories, interviews, secondary information and various databases.

The second stage involved the development of a multimodal transportation and logistics model that comprises Panama´s intermodal cargo flows data from 2006-2015. It also includes a costs and time model for the main logistics chains analyzed. This component captures the key elements that comprise the relevant logistics costs associated with the transport of each product segment through the study area on the main transportation modes. The model also includes a market demand module that summarizes a demand forecast for the major market segments over the time period of the analysis. This analysis is based primarily on the assessment of a base year demand, which is derived from data obtained from each of the modal options.

Demand is organized by product segment, zone of origin and destination. For future forecast years, demand matrices are estimated based on establishing causal relationships between base demand and key "drivers" of demand. This allowed the identification of future transportation infrastructure needs, as well as the actions required in order to attract potential market segments that could benefit from VALS. Finally, this analysis, along with the benchmarking of seven international logistics hubs, were utilized to establish five strategies, based on international best practices, that support the action plan.

Along with the action plan, a digital elevation model aided in the identification and dimensioning of areas with high potential for logistic developments. Additionally, an inter-institutional cooperation scheme analysis based on international case studies and local consultation was developed.

This national integrated initiative provides defined actions for the short, the medium and the long-term planning of transportation infrastructure and logistics. This initiative is based on quantitative and qualitative analyses, with the purpose of meeting future needs of both logistics projects and cargo demand flows by taking a proactive role. The recommendations contained in the PMZIC have been included as one of the pillars of Panama's National Logistics Strategy, recently launched by the Logistics Cabinet.

The remainder of this paper is organized as follows. Section 2 describes the methodology, the assessment of the hub and gap analysis based on the current state of the hub. Section 3 contains a short explanation of the development of the multimodal transportation and logistics model and main outputs. Section 4 describes the five strategies obtained for the development of the hub based on the benchmarking of seven top logistic hubs. Section 5 details the short, medium and long term action plans for the development of VALS and cargo and transportation infrastructure supported by the model results and five strategies. Section 6 contains a summary of the proposed implementation aspects that resulted from the case study analyses. Finally, Section 7 presents some final notes.

---

[3] Administrative Unit of Reverted properties of the Panama Canal (UABR) was established in 2007 as a replacement for the former Inter-Oceanic Region Authority (ARI). This Unit is in charge of managing many of the buildings and areas belonging to former US military bases.

[4] Panama´s Housing Ministry

[5] Ministry of Commerce and Industries

[6] Ministry of Economy and Finance

2. METHODOLOGY, CURRENT STATE OF THE HUB AND GAP ANALYSIS

2.1 Methodology

In May, 2016, the Panama Canal the Authority awarded by public tender a one year contract for the development of the Transportation Infrastructure and Cargo Logistics Master Plan for the Interoceanic Zone of the Panama Canal to the company Louis Berger S.A.S. The study comprised 23 deliverables among workshops, reports, GIS models and a cost and multimodal logistic Excel model.

The study area initially considered the interoceanic zone (ZIC) and encompassed the administrative units of the provinces of Panama, Panama West and Colón; and the Districts of Panamá, Colón, San Miguelito, Arraiján, Balboa and la Chorrera, from Vacamonte onthe West side, to Tocumen on the East side.

The preliminary base study area was initially 149,602 hectares. Protected areas and watershed comprised 33 percent of the total area .By excluding these protected areas, the ZIC study area covered 75,754-hectare of base study area. Figure 1 shows the preliminary ZIC study area boundaries and the protected areas and water bodies.

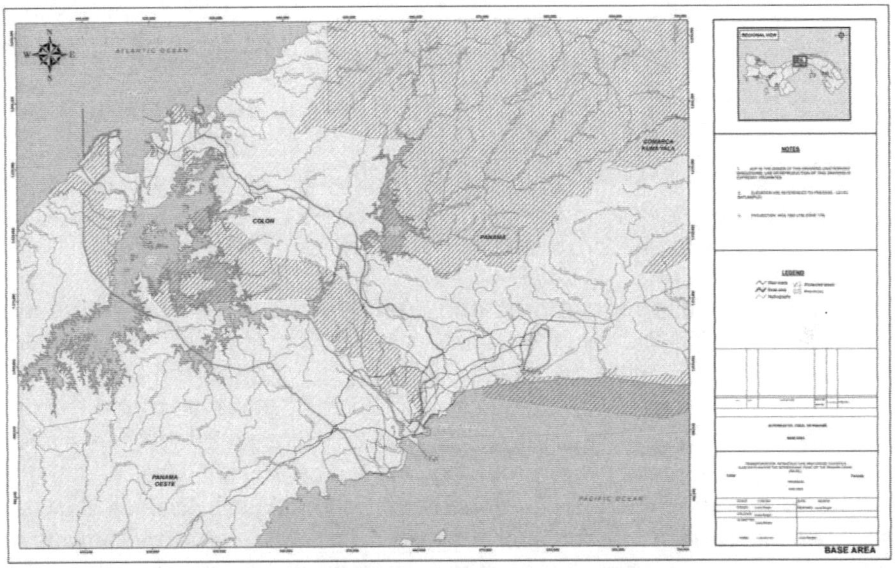

Figure 1. PMZIC preliminary study area

Four main blocks of analysis composed the PM-ZIC. The first part comprised a current situation assessment. Data collected from interviews, secondary information and databases from internal and external parties was used to build a Transportation Infrastructure Inventory in the ACP ARCGIS platform and a Panama International Cargo Flows Database in Excel.

- **Physical and digital inventory of infrastructure, transportation services and projects, Current Cargo Flows Inventory:**
  - Road network inventory: Includes 915 km of roads, with service levels, pavement conditions, and ongoing road projects with impact on the logistic platform of the ZIC. This inventory includes the results from the "PIMUS II[7]" study carried out by Metro de Panamá.
  - Railway network inventory: 72 km of railway included, equipment inventory and service levels obtained by field observation and interviews.
  - Ports inventory: ten international seaports classified by cargo type, infrastructure capacity (berths and yards), and service levels. Main data of port movements was obtained from

[7] Plan for an Integral and Sustainable Urban Mobility by Metro de Panamá

Panama Maritime Authority Statistics Department. Infrastructure inventory from field observation, interviews and aerial photography, among other methods.

- Airports inventory: includes four international airports with current and projected capacity. Information was obtained from field observation and interviews with the Civil Aeronautics Authority and Tocumen S.A.
- Logistics Infrastructureinventory: includes 13 free zones, 18 logistics zones, 2 special economic zones already stablished and 10 projects; their total area in hectares and current development level.

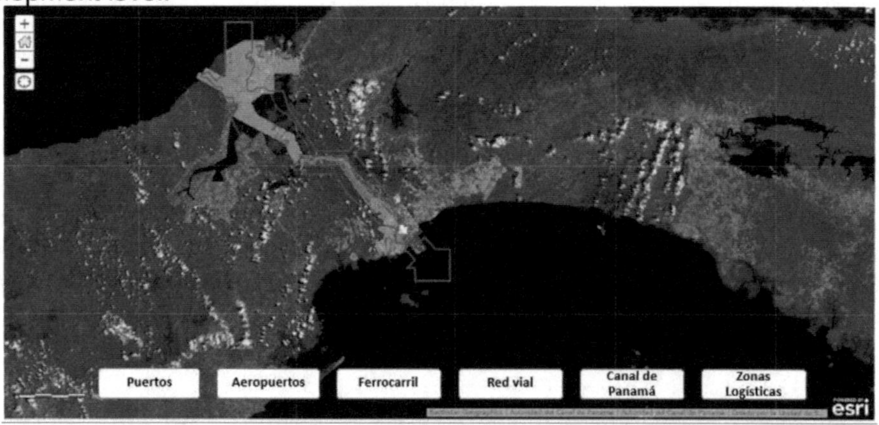

Figure 2. ACP´s GIS containing the ZIC Transportation Infrastructure Inventory

- Panama Canal: Panama Canal cargo traffic in tons by market segment and the Technology Information Division ESRI inventories and GIS platform. Land Use affairs by interviews with the Corporate Communications Vice-presidency.

- **Current logistic chains mapping and analysis and Potential Chains identification**
  A selection of logistic chains was conducted taking into account the relevance of the total volume and value in the external trade, the actual VALS taking place in the ZIC, a qualitative assessment of the level of employment and wealth generation and the priority development areas for the country. A qualitative BCG[8] Matrix of current segments and VALS in Panama was conducted to select the target logistic chains for analysis. A focus group and several interviews were held to aid in the mapping of each logistic chain and collecting the inputs for the time and cost model for each of them. The mapping and analysis was conducted for perishable products, shoes and clothing, electric and electronics, cars, pharmaceuticals and alcoholic drinks.

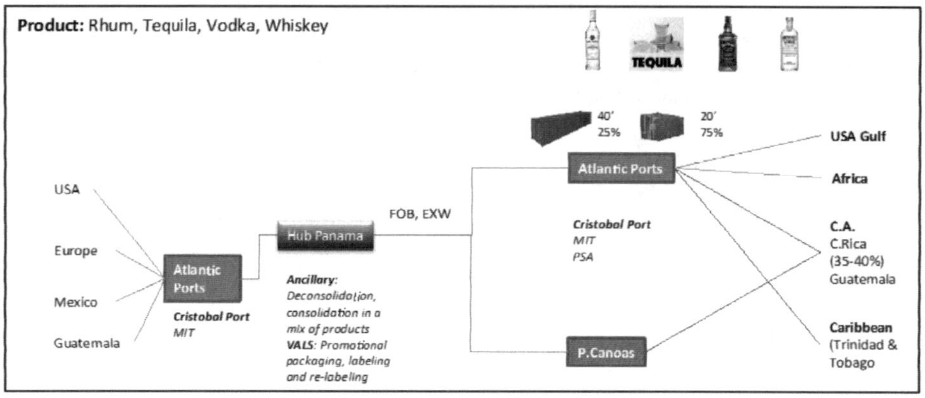

[8] Help a business consider growth opportunities by reviewing its portfolio of products to decide where to invest, to discontinue or develop products. It's also known as the Growth/Share Matrix.

Figure 3. Example of the mapping of current logistic chains for vehicles and APE

From the mapping for the selected chains it was possible to identify most of the aspects hindering the performance of each specific chain, and with the time and cost model it was possible to simulate the potential savings by reducing these inefficiencies.

In the example of the alcoholic drinks imported by sea from Europe, processed in the Colon Free Zone and then exported to the Caribbean an optimized cost and time saving could be up to 36% and 8%, respectively.

For the identification of potential logistics segments a market research study was conducted considering: a) value of the good; b) logistics costs as compared with neighbors and competitors; c) volume and density of value of the good after providing VALS; and d) vertical integration for a given market and possibility to change the structure of the logistics chain. Based on this and by consolidating the segments identified through the analysis of external trade and transshipment patterns, as well as those derived from exogenous factors, a set of 3 logistics segments with 8 logistics families has been identified. The market analysis included interviews with public and private institutions. These provided an understanding of current logistics patterns and Panama's relative attractiveness for these patterns. Interviews included associations of exporters, exporters of perishable goods, investment promotion agencies, institutional logistics operators, and logistics operators (firms).

- **Benchmarking of main international logistic hubs**
  Seven international value added logistics hubs were evaluated. These were Shanghai, Hong Kong, Singapore, Dubai-Jebel Ali, Rotterdam, Philadelphia and Houston. The assessment allowed the identification of exogenous factors as well as strategic capabilities critical to develop a VALS hub: Threshold resources, critical success factors and unique resources.
- **Development of the multimodal transportation and logistics model**
  This fully integrated model includes three main components, described as follows:
  - ***Cost and Time Model***. This component captures all the key elements that comprise the relevant logistics cost associated with the transport of each product segment through the study area on the major transportation modes. The analysis focuses on the range of mode and route options that are available to goods flowing between the origins and destinations that are defined as relevant. The model focuses on modeling all-in logistics cost as perceived by decision makers who make route decisions. Critical elements that are included are transportation costs for each mode, the perceived value of time for each product segment, transshipment, handling and storage costs, among others.
  - ***Market Demand Model***. This component summarizes the forecast of demand for the major market segments over the period of the analysis. This is based primarily on an assessment of base year demand which is derived from data obtained from each of the modal options. Demand is organized by product segment and zone of origin and destination. For future forecast years, demand matrices are estimated based on establishing causal relationships between base demand and key "drivers" of demand. Potential drivers include mainly macroeconomic growth rates by country as a qualitative assessment of potential cargo flow patterns through the region, IHS projections for containerized imports and exports for LAC were used for the short term unrestricted transshipment forecast. For the long term forecast, GDP was used as a base, utilizing IHS projections for elasticity. A combined analysis was applied in order to evaluate three potential scenarios for cargo flows within the hub.
  - ***Demand Assignment Model***. In this final component, the demand is firstly adjusted using market capture scenarios and the allocated among mode options. In the case of the ZIC area, most of the cargo (Re-exports and Transshipment) are originated from and are in destination to outside the ZIC. Market capture factors such as market trends and external competition have been considered to build market capture scenario. In case of internal competition within the ZIC area (which is the case for bi-oceanic transshipment), the assignment has been done using a function of relevant logistics cost of each according to

a logit distribution function. This function estimates the market share of demand captured by each potential mode in the ZIC in relation to other mode. This function is calibrated to reproduce the mode shares observed during the base forecast year, and measure changes in share captured in future years as a function of changes in the relative costs.

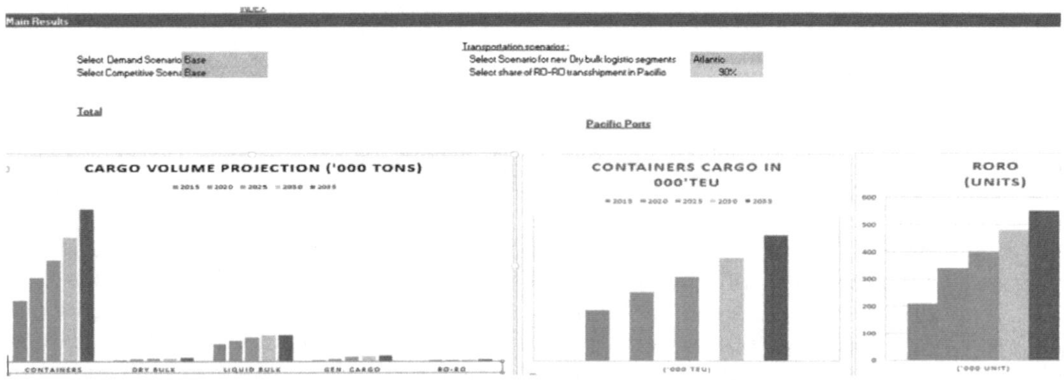

Figure 4. Results tab of the Excel Multimodal Transport Model

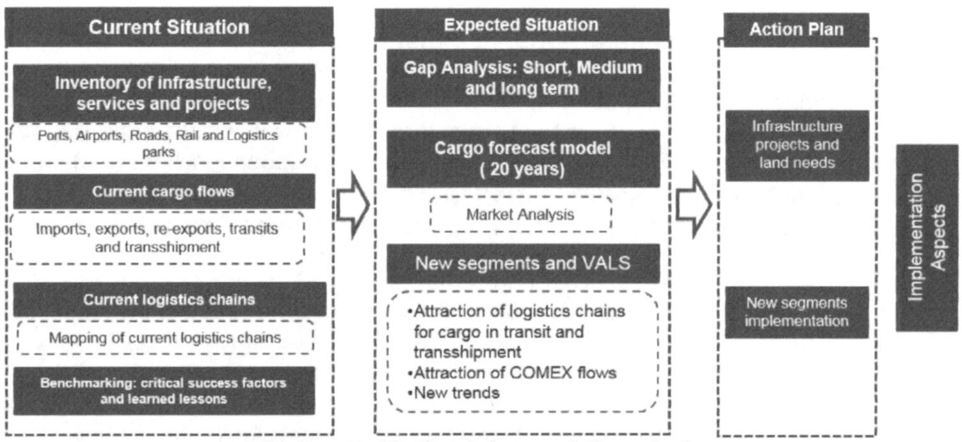

Figure 5. Methodology of the study

During the formulation, the model was continuously validated through workshops, focus groups, direct and indirect interviews and field surveys with stakeholders

Figure 4. Meetings carried out with key stakeholders

3. FIVE STRATEGIES FOR THE DEVELOPMENT OF THE HUB

The benchmark of international logistic hubs provided some key findings regarding needs and opportunities for the different market segments that could benefit from operations in Panama. Most of these market segments share critical needs in terms of quality and efficiency of the existing infrastructure, services and control processes. In order to address these needs, five strategies based on international practices have been decided to serve as a guideline for the development of the Panamanian VALS Hub.

Panama can consolidate and expand its current service and market by:

- **Generating a bidirectional Hub-and-spoke system**
  Panama can play a role in the optimization of current logistics patterns for several chains in the Region. Current chains using Panama share a similar pattern. Goods come to Panama for the distribution to the Region. Export concentration patterns were only identified in perishable goods and beverages. Perishables do not receive VALS but only air-air and road-air transshipment. Beverages go to the Region, however a very small volume goes to the US and Africa. The proposal consists of creating a bidirectional hub-and-spoke system by expanding current regional distribution of extra-regional goods with consolidation of intraregional and extra-regional exports.

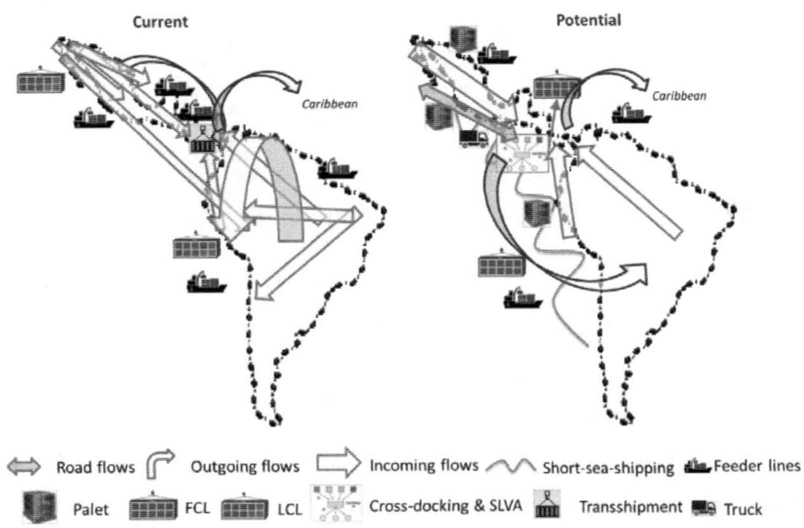

Figure 6. Current and potential flows for Pharmaceuticals, Home Appliances, and Automobiles & Spare Parts

- **Three logistics poles definition**
  Three key logistic poles have been identified in Panama. Consolidating and further developing these poles will generate economies of scale and agglomeration, improve transportation infrastructure investments, homogenize special economic regimes, and allow the integration of processes among the poles, and help coordinate the scheduling of the development program in accordance with the expected demand.

- **Multimodal System**
  A need for an efficient multimodal system with highly efficient fiscal priority corridors with real-time tracking of vehicles in a digital optimization platform has been identified.

In regards to high standard logistics infrastructure, new specialized logistics activity zones for perishables and dry bulks will be required. Tocumen Free Trade Zone and the Panama Canal logistic zone are new core components of the logistics supply and will be important assets in the near future.

An integrated port zone is part of the proposal, and consists of the operational integration of the ports located at the ZIC in order to optimize existing assets. It strengthens Panama's unique resources: its inter-ocean access and the inter-terminal transshipment. A booking system linked to the Panama Logistics and External Trade Platform and its PCS (Port Community System) functionalities, will be an important component.

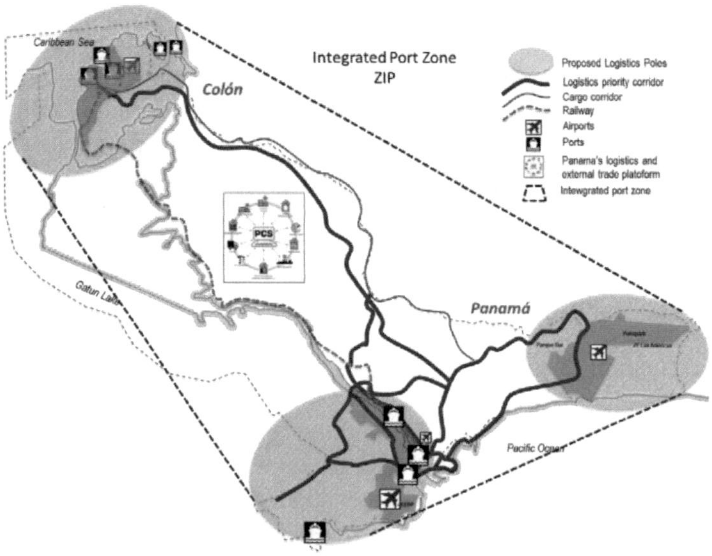

Figure 7. Highly efficient multimodal system proposal and three logistic poles

- **New logistics segments and VALS**

The current Panamanian supply of logistic services is mostly composed of ancillary services – or support services related to transshipment and consolidation flows. These are mostly provided in ports and airports.

VALS are provided to goods in transit entering ZLC, ZEEPP and FTZ and destined to re-exportation. The broadest range of services corresponds to ZLC – also the main contributor to re-exportation flows. However, current VALS are the lowest in terms of value-addition. High-value added logistic services such as assembling and postponement are rather scarce, and reverse logistics, quality control and reconditioning almost inexistent. Pharmaceuticals is the logistics segment that most contributes to value-addition in Panama. In 2015 the average density of incoming goods was 66,000 USD/Ton vs. an outgoing of 107k USD/Ton, a 61% increase; beverages followed with a 24% increase. For the highest-volume segments the change range from 2 to 12%. The proposed strategy is to maintain or improve the current market share of the analyzed existing segments and VALS and generate loyalty by developing new VALS and services.

Three new segments identified as target markets are Perishable goods, B2C (business to consumer segments) and Dry bulks. Logistics chains to be served differ depending on the logistics family.

The perishables segments looks after capturing traditional exports aiming at reaching new extra-regional markets (namely Asian countries) for non-traditional goods usually highly-perishable goods with highly diversified requirements of controlled atmosphere. Some of the VALS offered can include consolidation, cross-docking, cold chain traceability and intermodal services. B2C segments look after deepening the current offer to more value-added VALS and to develop the e-market channel. It ranges from apparel, shoes and textiles to pharmaceutical, medical equipment and vehicles and spare parts, among others. VALS targeted to dry bulks, are mainly proposed to the agriculture related bulks (grains and fertilizers) for packaging, storage, unitization and consolidation and distribution in containers.

- **100% digital hub**

  Logistics optimization will rely on a national Cargo Community System (CCS) that will link all transport and logistics infrastructure. Logistics priority corridors will be also linked to the CCS in order to efficiently implement their fiscal corridor functions. This integration will benefit the private sector by providing them information on real-time flows that can be used to optimize existing assets.

  The AIG (Government Innovation Agency), which leads the technological adaptation on e-government nationwide, along with the Logistics Cabinet, is currently developing the project "Technological Platform of Logistics and External Trade Integration", a platform for logistics processes and operations digital integration.

  It will allow implementing an integrated single window for all administrative procedures and it is expected that the platform will link all stakeholders related to logistics processes in order to optimize the information flow.

Figure 8. Proposed cargo community platform

4. SHORT, MEDIUM AND LONG TERM ACTION PLANS

Short, medium and long-term action plans are proposed for transportation infrastructure, services development, and processes reform based on the five strategies, the long term demand forecast and land availability analysis. The action plans look into overcoming capacity constraints of transportation infrastructure to serve current and future demand by existing and new logistic segments for the next 20 years. It also includes the actions to be taken in order to ensure a gradual, stable and sustainable supply of logistic services, as well as to generate a sound value proposition of VALS to new target segments. It also includes a program for land development (or poles development) according the expected demand phased for up to 20 years and a proposal for reserved areas after 2035. The land use forecast is integrated to the ZICs Multimodal and Transportation Infrastructure Model and calculates the area based on a Ratio (ton/ha) Ratio based on ZLC development Standard.

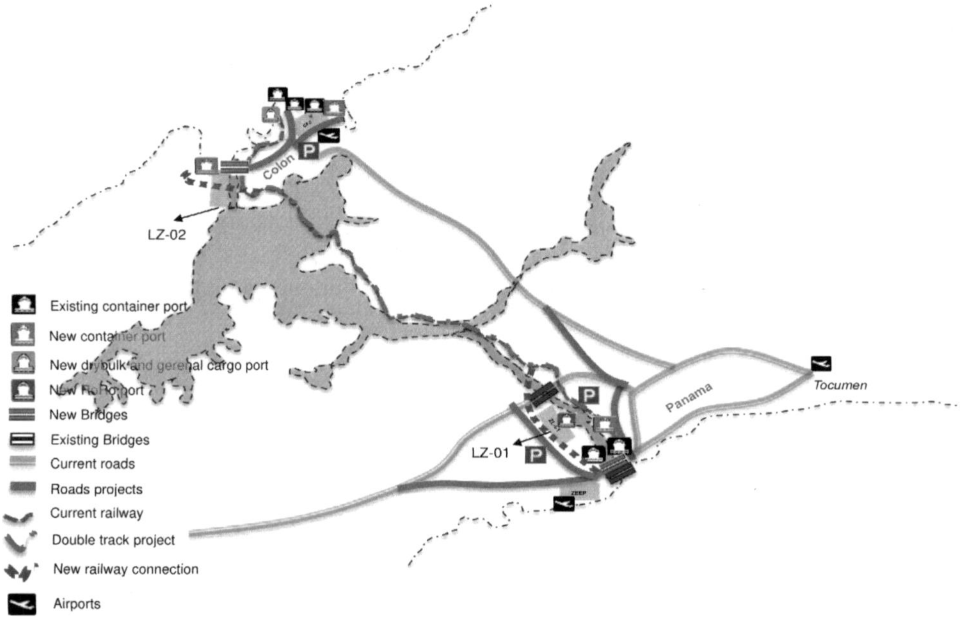

Figure 9. Logistics areas development suggested by the Transportation Infrastructure Action Plan

5. IMPLEMENTATION ASPECTS

From a case study analysis and best practices that could be applicable to the ZIC, cooperation schemes are proposed at two levels: strategic and operational. The first one is long-term oriented and look after ensuring sustainability and optimizing the execution of the PM-ZIC as a whole, whereas the second points out at beginning the execution of the plan.

- Strategic Level: Institutional, legal and financial components that need to be adjusted in order to streamline the execution of mega-infrastructure projects and effectively launch VALs to new logistics segments. From the institutional perspective, some implemented figures consider a coordination agency focused on financing and capacity building, reinforcing capabilities to capture and assign financing resources on a long-term basis and provide technical assistance to all transversal institutions involved in the execution of the PM-ZIC in all its phases. At the same time, a short-term promoting agency to manage complexities of executing mega-infrastructure projects could speed-up the process.

A legal framework for standardising the different existing regimes within the ZIC (3 poles), and a possibility of creating partial plans in the short term, determining the pole´s borders with its corresponding land management guidelines.

In regards to the financial resources, pre investment funds allocation on a continuous basis and the creation of a solid PPP mechanism that contributes along with public resources and loans.

- Operational level: Developing business plans, creating strategic partnerships, launching a massive training program of HR at all levels, capacity building as well as implementing a regulatory scheme for the ZIP are all actions that should be faced in the short time.

6. CONCLUSIONS

- The need for this master plan was identified in 2015 ´by the Logistics Cabinet´s Infrastructure Committee, which is a part of the Logistics and Competitiveness Secretariat of the Ministry of the Presidency. This committee is currently headed by the Panama Canal Authority and is composed of public institutions and logistics representatives from the private sector. The document was developed jointly with the National Logistics Strategy, led by the Logistics Cabinet.
- This Master Plan is the first integrated transportation infrastructure and cargo logistics planning effort in the country. The study identifies the main challenges for capitalizing on surrounding areas of the Panama Canal and proposes a set of action in infrastructure, services and implementation aspects to offer an efficient platform for value added logistics to contribute to the Canal´s and conglomerate land profitability.
- In May, 2016, the Panama Canal the Authority awarded by public tender a one year contract for the development of the Transportation Infrastructure and Cargo Logistics Master Plan for the Interoceanic Zone of the Panama Canal to the company Louis Berger S.A.S. The study comprised 23 deliverables among workshops, reports, GIS models and a cost and multimodal logistic Excel model.
- Most of the existing market segments share critical needs in terms of quality and efficiency of the existing infrastructure, services and control processes. Based on the benchmarking, the lesson is that Panama could consolidate and expand its current markets by investing in infrastructure and transport services to expand the hinterland. At the same time, generating a bidirectional Hub-and-spoke system, intra and interregional, implementing world-class logistics zones consolidated in three logistic poles to generate economies of scale and mitigate externalities, consolidating a fully intermodal system, with well differentiated short and long distance trips, reducing OPEX by tackling all existing operational constraints, generating high-quality IT infrastructure and implementing a 100% digitalized system.
- Intraregional trade it mostly composed of fast-consuming and industrial goods, in particular in Central America; industrial inputs, food preparations for human and animals, minerals, petroleum, among others. These flows do not usually require VALS rendered in a logistics hub. Durable consumer goods on the other hand, can benefit from VALS in a logistics hub. Panama has a wide experience with durable goods such as clothing, footwear, cosmetics, vehicles and pharmaceuticals, among others; therefore, the challenge is in expanding to a new segments and diversifying the nature of VALS for these intraregional flows.
- The development of a VALS Hub is a national priority aimed at promoting the exportation of services, value capture, and therefore maximizing the opportunities for employment generation.
- The PM-ZIC is a complex endeavor that implies public and private investments in a relatively short period, thus demanding that the whole set of public and private resources be available accordingly in an optimal manner.
- The execution of the plan requires an active involvement of public and private institutions to finance, promote, and execute its diverse components at strategic and operational levels.

References

Panama Canal Authority (2017) Transportation Infrastructure and Cargo Logistics Master Plan for the Interoceanic Zone of the Panama Canal, Panama Canal Authority, Panamá.

Panama Canal Authority (2017) Executive Summary of Transportation Infrastructure and Cargo Logistics Master Plan for the Interoceanic Zone of the Panama Canal, Panama Canal Authority, Panamá.

Panama Canal Authority (2005) ACP land use plan. URL:micanaldepanama.com/wp-content/uploads/ 2012/08/ Normas-de-Iso-TipoII.pdf

Existing Laws for National Parks in Panama

The Republic of Panama's "Unidad Administrativa de Bienes Revertidos" (UABR) land use plan for the area is included as files "Mapa1-Plan Regional de Usos del Suelo y los Recursos Naturales de la Región Intreoceánica.pdf" and "Mapa2-Plan General de Usos del Suelo del Area del Canal.pdf".

Land Use Plan (MIVIOT) (original text in Spanish).

The Canal's Impact on the Panamanian Economy (May, 2012)

Dimension Program of the Panama Canal´s Impact over the country´s economy Nathan Associates Inc. for the Panama Canal Authority (May, 2012), (text available in English).

Panama Canal Master Plan (2005-2025) dated (June, 2006)

The Economic Impact of the Panama Canal on the Republic of Panama - Executive Summary (May 2006). (URL: http://micanaldepanama.com/plan-maestro-2005-2025/plan-de-estudios/financieros-y-economicos/estudio-de-impacto-economico-del-canal-en-el-ambito-nacional/).

Government Strategic Plans (2010-14) (2015-19) (original text in Spanish).

Tocumen S.A. (2016) Master Plan for the Development of Tocumen International Airport, Panama Pacífico Airport and Cargo Master Plan for Tocumen Internatioanl Airport.Tocumen S.A., Panama City, Panama.

Inter-American Development Bank (2014). Cargo Logistics National Plan (PNLOG) 2014-2024. (original text in Spanish). IDB, Panama.

Metro de Panama S.A. and IDB (2015) Metropolitan Area Urban Mobility Plan (PIMUS) I and II. Panama City, Panama.

IDB (2012) Panama Foreign Trade Process Mapping and Diagnosis. Studies developed for the Inter-American Development Bank (original text in Spanish).

IDB (2013) Analysis, strategy and tools for cargo logistics and Mesoamerica´s foreign trade improvement. ALG for the Inter-American Development Bank.

Panama´s Air Logistics Strategy. ALG and Webber Air Cargo (original text in Spanish)

Cargo Flows and Geographic Information System land use layers from Governmental databases (AMP, ANA, Panama Canal Market Segments, Civil Aeronautics Authority, and Municipality of Panama, among others).

LOGISTICS & INFRASTRUCTURE: PROGRAM MANAGEMENT IN THE PANAMA CANAL EXPANSION

by

Richard Robertson, P.E.,[1] Ilya R. Espino de Marotta, P.E.,[2] and Ilona Modly Hogan, J.D.[3]

Abstract - In 2006, the citizens of Panama voted to approve a national referendum for the **Panama Canal Expansion Program (PCEP)**. The goal of the program was to grow the global shipping industry infrastructure and expand the economic opportunities of Panama.

Program Management Definition - Program Management is the delivery of multiple, interrelated projects in a coordinated process for the benefit of the owning entity.

Autoridad del Canal de Panamá's (ACP's) Consideration of Program Management Models - Owner Management Model, At Risk Model, Consultant Management Model, Owner's Agent Model, Integrated Program Management Model

ACP Selection - **Integrated Program Management** and Program Management Services Contract Awarded to CH2M in August 2007

One Team One Mission Organization & Collaboration - Leveraging a cross-cutting approach from the program management development of tools and processes to establish the governance of the PCEP to a mirror manager approach of CH2M, with ACP counterparts, offering guidance, advice, and support for ACP's authority and decisions

PCEP Stakeholders: People of Panama, Autoridad del Canal de Panamá (ACP), Board of Directors and Global Maritime Industry

Risk Management - A key element of any Program Management model is the allocation of risks and responsibilities. Risk transfer works best when you get as close as possible to a "win-win" scenario. A risk profile with an imbalance inevitably leads to the detriment of one or more parties.

In the case of the PCEP, the integrated ACP-CH2M team identified the risks during the tendering process and the risk allocation was improved, in favor of the tenderers, as a result of open dialogue. This risk approach was maintained throughout the execution.

Relevance Statement - The relevance of this presentation can be generalized for future large infrastructure initiatives. For the Panama Canal Expansion Project, there were two main topics:

- Integrated Program Management Model selection and implementation
- Management of Program Risks (Contracting, Delivery and Reputational Risks)

Keywords – Panama Canal, Program Management, Infrastructure, Megaprojects, Collaboration

[1] CH2M Hill Panama (now Jacobs), Panama Canal Expansion Program, Rick.Robertson@ch2m.com

[2] Autoridad del Canal de Panamá, imarotta@pancanal.com

[3] CH2M Hill Panama (now Jacobs), Panama Canal Expansion Program, Ilona.Hogan@ch2m.com; the authors thank the work of former CH2M President of the Transportation Business Group Terry Ruhl and his PMI colleagues for their early contributions to the process, the program and project management differentiators on the PCEP.

In 2006, the citizens of Panama voted to approve a national referendum for the Panama Canal Expansion Program (PCEP). The goal of the program was to grow the global shipping industry infrastructure and expand the economic opportunities of Panama. This program came after years of studies performed by the Autoridad del Canal de Panamá (ACP) in completing a Master Plan for the development, growth and improvement of the Canal with a 25-year outlook. This would be accomplished through an additional expanded size lane of traffic — via a new third set of locks — to double the Canal's tonnage capacity and allow the transit of much longer, wider ships through the existing canal waterway. When the expansion locks were inaugurated in June 2016 they increased ship throughput capacity to 18,000 vessels with an increase in capacity to approximately 14,000 TEUs, a standard unit for describing a ship's cargo carrying capacity or a shipping terminal's cargo handling capacity. See Figure 1 below.

Figure 1: Neo-Panamax Mega-container traverses the Panama Canal Expansion Project in June 2016

During the development of the PCEP concept, it became evident that the **Autoridad del Canal de Panamá (ACP)** could benefit from the services of a Program Management firm due to the program's risks in logistical and technical complexity; the interrelated multiple contracts; and the multi-billion-dollar budget which encompassed resource needs beyond the ACP's existing engineering and procurement support managing the operations and maintenance and routine capital improvement projects. Therefore, a decision was made to partner with an experienced Program Management firm to advise and assist throughout the execution of the expansion program.

Another key decision made by ACP was to continue the historic use of English as the contract/program language. ACP felt this would encourage bidding by a wider range of international contractors for the various PCEP contracts. At the time of the procurement, the top Program Management firms were U.S. firms, as ranked by *Engineering News-Record (ENR)*. Notwithstanding this selection of a common contract language, in practice the cohesiveness of project execution was affected by the fact that the four **Grupo Unidos por el Canal (GUPC -** Third Set of Locks' design-build contractor) shareholders are Spain's Sacyr Vallehermoso, Italy's Salini-Impregilo, Belgium's Jan de Nul and Panama's Constructora Urbana S.A. GUPC partners represented four countries and three languages. These linguistic distinctions manifested themselves as the Pacific Site spoke English under the leadership of GUPC partner Impregilo from Italy while the Atlantic site retained the Spanish language under the leadership of GUPC partner Sacyr from Spain.

Program Management Definition

Program Management is the delivery of multiple, interrelated projects in a coordinated process for the benefit of the owning entity. The need is often associated with a spending growth for capital infrastructure programs reflecting increased risks and complexity arising from the interfaces between multiple projects as well as the effects of varying project schedules. Program durations are longer than individual projects because programs encompass multiple projects and uncertainty is likely to be higher because of the likelihood of encountering environmental, financial, political, technical solutions and other changes.

Key Indicators that identify when Program Management is needed:

- Complex or multiple projects
- Increased capital spending required
- Limited in-house client resources and expertise
- Many projects of multi-year duration
- Schedule pressure
- Project delivery organization is not permanent
- Single point of responsibility is desired
- Goals are clear; delivery approach is not

Why Program Management?

Program management is the coordination of Resources to plan and deliver a collection of related projects to obtain benefits that would not be attainable if managed separately. Complex programs have traditionally been implemented using standard project management methods. As programs have continued to grow in risk and complexity, a more disciplined overall delivery method was developed—program management. With multiple ongoing projects, resources such as key staff, technical expertise, managers, electronic systems, buildings/facilities, tools, equipment, etc. are stretched and often duplicated to meet delivery demands. Program management optimizes resources to deliver benefits not attainable using the traditional project delivery model. See Figure 2 below.

While technical knowledge and experience are critical in all work, providing the best delivery method is equally critical—enabling each project to meet client goals and to be completed within budget and schedule.

Mentorship and career growth are essential for the industry and for clients – this is more readily available in a shared working environment with opportunities and empowerment of staff leveraging their experience.

Management of the interfaces between projects reduces program complexity. Every project and client has a unique set of delivery requirements that must be understood to develop efficient and effective solutions.

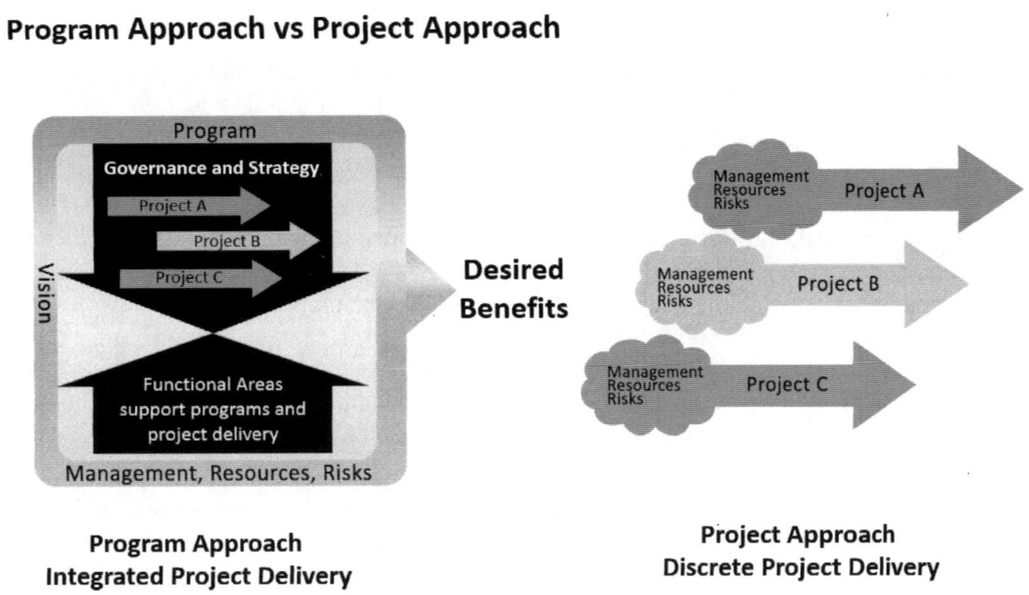

Figure 2: Program vs. Project Approach

Through extensive experience and lessons learned, we know that no two programs are alike and no model is right or wrong. Identifying the client's unique vision, goals and desired benefits are key to the program's success.

Program Management Framework and Best Practices

Four vital quadrants define Program Management Best Practices as in Figure 3 below:

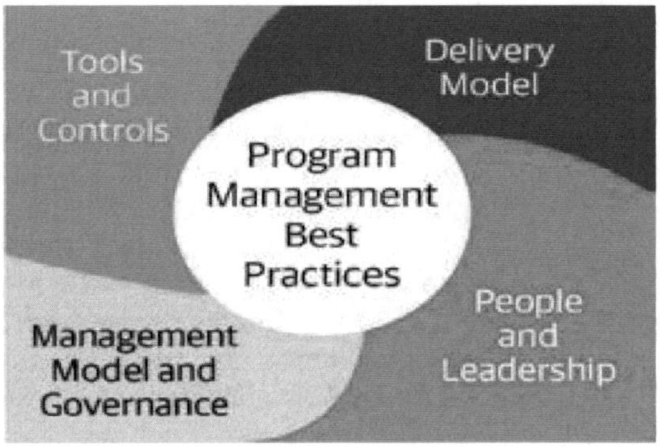

Figure 3: Program Management Best Practices

The **Program Management Framework** then organizes and integrates these quadrants to optimize a value proposition based on lessons learned and best practices. A scalable Program Management Framework standardizes delivery strategy, processes, tools and resources around a common platform to successfully deliver programs tailored to clients' vision and to integrate seamlessly with their existing systems. The Framework includes the following four integrally related components which provide the tools and control for efficient program delivery throughout all phases from the early concepts, during the initiation, and throughout the execution and closeout.

Program Initiation provides structured processes and tools for rapid client and program management team integration, allowing the PM team to focus on the client's goals and critical activities, while providing cost-effective, collaborative delivery practices. During Initiation, the Program Leadership is formed and the Management Model and Governance developed.

Program Delivery Guidelines provide an electronic Program Management Manual for implementing program-specific processes and procedures and applying the critical thinking to adapt the procedures to meet the program-specific requirements. The Platform documents the Delivery Model and related Tools and Controls for consistent delivery.

Program Performance Tools that allow program-specific information management for consistent critical data and program progress reporting, including a web-based dashboard. This transparent access to information enhances delivery efficiencies, program quality and decision making.

Program Maturity - Assessment tools are used to periodically evaluate the developmental status and operational effectiveness of the program's governance and infrastructure. This can be a matrix, spreadsheet, scorecard, etc.

Critical success factors include schedule, lowest overall cost, scope clarity, staff development, owner control and risk tolerance. Prioritizing and understanding these factors by use of several diagnostic tools, results in a program delivery model that best fits the client's needs and situation.

Program Management Economics

Before considering Program Management concepts, it is helpful to examine the economic benefits of a well-planned and well-executed program model. Various studies illustrate the value of Program Management on large complex projects and even on the smaller projects that form a Program Management portfolio. Each program is unique with its mix of social, economic and political complexity, but a key success factor is defining the program risks early. With little room for error in today's business environment, the proper identification and mitigation of risks as part of the overall Program Management process can serve to minimize cost and schedule overruns, assuming all stakeholders are aligned around a program's goals and objectives.

A direct correlation between staffing a PMO correctly and its ultimate success can be observed. In its *Pulse of the Profession™ In-Depth Report: The Impact of PMOs on Strategy Implementation*, the Project Management Institute reports that high-performing PMOs – those that complete 80 percent or more of projects on time, on budget and meeting original goals – are more than twice as likely to have the right skills base in regard to project management experience and training.[4]

In PMI's 2015 *Pulse of the Profession®: Capturing the Value of Project Management* research revealed that "high-performing organizations are far more likely to focus on talent management, establishing ongoing training, and formal, effective knowledge transfer."[5] This is especially important in project and program management, where technical skills are enhanced by leadership and its strategic and business management capabilities that are nurtured through experience.

The size of the program team may also factor into the equation. Generalizations can be hard to make given that each program is unique in its size, scale and complexity. However, after a cursory review of some successful major infrastructure programs around the world, as compared to the total installed cost of the program being constructed, the size of the Program Management team (owner plus consultants) tends to hover between US $15 million to US $50 million of installed cost. More research is needed in this area in order to draw more accurate correlations around optimum staffing.

Program Management Models: Seeking the Right Balance

The ACP evaluated the various models of Program Management with a goal of selecting a working construct that aligned with its short- and long-term goals as well as its organizational and political objectives. The decision that fit the Panama Canal's Administration needs was that of an integrated Management team.

In addition, prior to awarding the Program Management Services contract, the ACP released a Request for Proposal for Program Management and Construction Management training, seeking a major university accredited as a Registered Education Provider by the Project Management Institute (PMI®). The University of Texas at Dallas was awarded the contract, and subcontracted CH2M, as the top-ranked (by *Engineering News-Record*) Program Management firm, to provide instructors. The Panama Canal certified 75 Program Managers under the PMP certification, also provided formal training to over 300 employees that were assigned to the Program.

[4] PMI's *Pulse of the Profession®* In-Depth Report: *The Impact of PMOs on Strategy Implementation (2013)*

[5] PMI's *Pulse of the Profession®* In-Depth Report: *Capturing the Value of Project Management* (2015)

Once the Program Management Services Contract was awarded, the ACP together and with the guidance and lead of CH2M developed a Program Management Plan for the Program and it was followed by over 1,200 ACP staff that were directly assigned to the Program.

Proyectos *JUNIO 2016*	Contratistas	Sub-contratistas	Total
Excavación del Cauce de Acceso Pacífico – CAP1	475	214	689
Excavación del Cauce de Acceso Pacífico – CAP2	492	620	1,112
Excavación del Cauce de Acceso Pacífico – CAP3	345	577	922
Ensanche y Profundización de las Bordadas del Norte del Lago Gatún	185	226	411
Entrada Norte del Cauce de Acceso del Pacífico	366	1,575	1,941
Dragado de la Entrada del Pacífico	817	520	1,337
Dragado de la Entrada del Atlántico	636	416	1,052
Excavación del Cauce de Acceso Pacífico – CAP4	2,069	1,047	3,116
Diseño y Construcción del Tercer Juego de Esclusas	14,580	9,872	24,452
Reforestación y rescate de vida silvestre	1,345	0	1,345
Alquiler de equipo pesado	244	0	244
Servicios paleontológicos y arqueológicos	85	0	85
Otros Contratos	2,835	338	3,173
Subtotal	24,482	15,405	39,887
ACP			1,281
Proyección del Total de Empleos Directos Acumulados			*41,168*

Figure 4: PCEP Staff Allocation by Project

ACP's Consideration of Program Management Models

With an accelerated understanding of the discipline, ACP evaluated the following models with the intention of selecting a working construct that aligned with its short- and long-term goals as well as it organizational and political objectives.

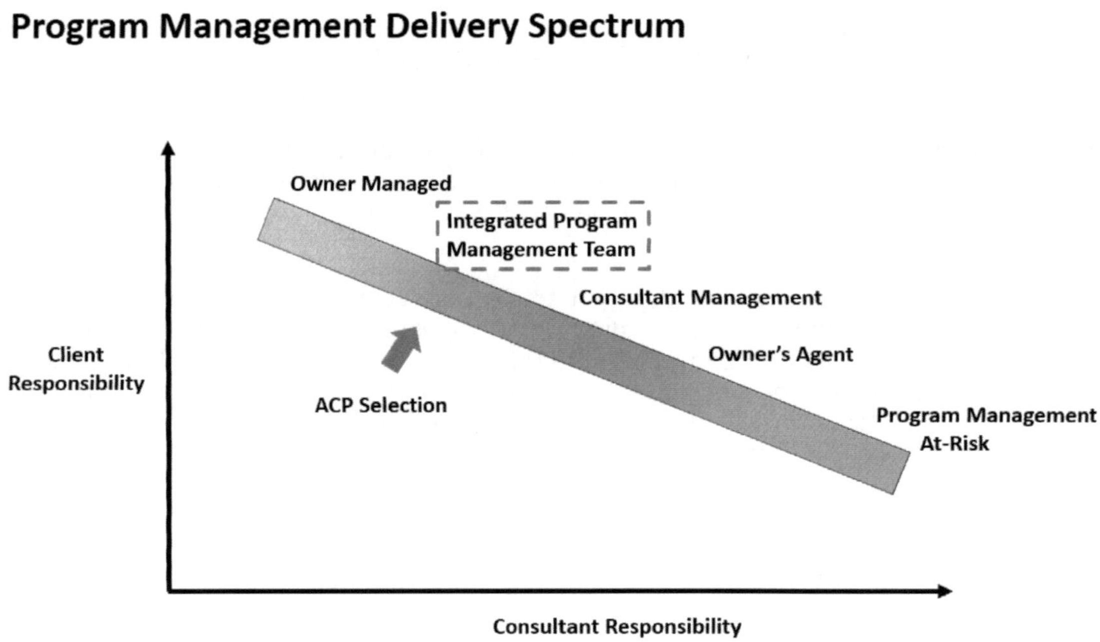

Figure 5: Spectrum of Program Management Models

Owner Managed Model: Owner provides project definition, hires the engineers and contractors and manages all contracts.

Integrated Program Management Model: Program Manager may be either the Owner or the Consultant, with staff from both organizations integrated at multiple levels. The Consultant staff may augment positions within the Owner's organization where expertise or experience is lacking.

Consultant Management Model: Consultant Manager may supplement the Owner's staff in an advisory role. The Owner hires the Program Manager, the designers and the contractors.

Owner's Agent Model: Owner hires the program manager and contractors, and may also in some cases hire the engineers, but the program manager is responsible for providing direct management of engineering services.

Program Management At Risk Model: Owner has less direct project input and control because it has assigned more risk to the Consultant and also assigns commensurate control over the Program to the Consultant.

There is more than one "right" program delivery method for a given program. And hybrids can be created when owners and consultants sit down to discuss mutual objectives. The key decision factor is for an owner to decide how much control it is willing to relinquish to a consultant and how a consultant wants to be compensated for assuming the risk.

ACP Selection: Integrated Program Management

The ACP selected the Integrated Program Management model in alignment with the following:

- ACP maintaining complete control of the program
- ACP remaining as a unified organizational entity but created a Vice presidency (Engineering and Program Management) that would be dedicated to the PCEP
- Developing within ACP the knowledge and skills for managing future large complex programs – with minimal, additional outside consultation

ACP's selection of the Integrated Program Management Model followed an industry trend over the last two decades away from independence (where the consultant would make many, if not all, decisions on behalf of the owner) to an integrated model (with the consultants acting in the role of seconded staff working on site with the owner), incorporating alternative project delivery methodologies.

CH2M was awarded the Program Management Services contract in August 2007:

- CH2M to conform with the ACP's organization by providing an integrated team to serve as program and construction advisors to the ACP;
- CH2M to train and mentor the ACP staff in program and construction management best practices, as advisors in mirrored roles, and to provide technology transfer;
- CH2M to develop program/project discipline, processes and procedures; and
- CH2M to assist with claim/dispute defense and claim/dispute mitigation.

With an integrated model, a significant amount of formal and informal learning takes place naturally as everyone works together, side-by-side. Under the terms of the agreement with the ACP and, as the work progressed, CH2M would phase out its staff, leaving in place an owner team trained in advanced Program Management and delivery and capable of operating and managing a world-class operation.

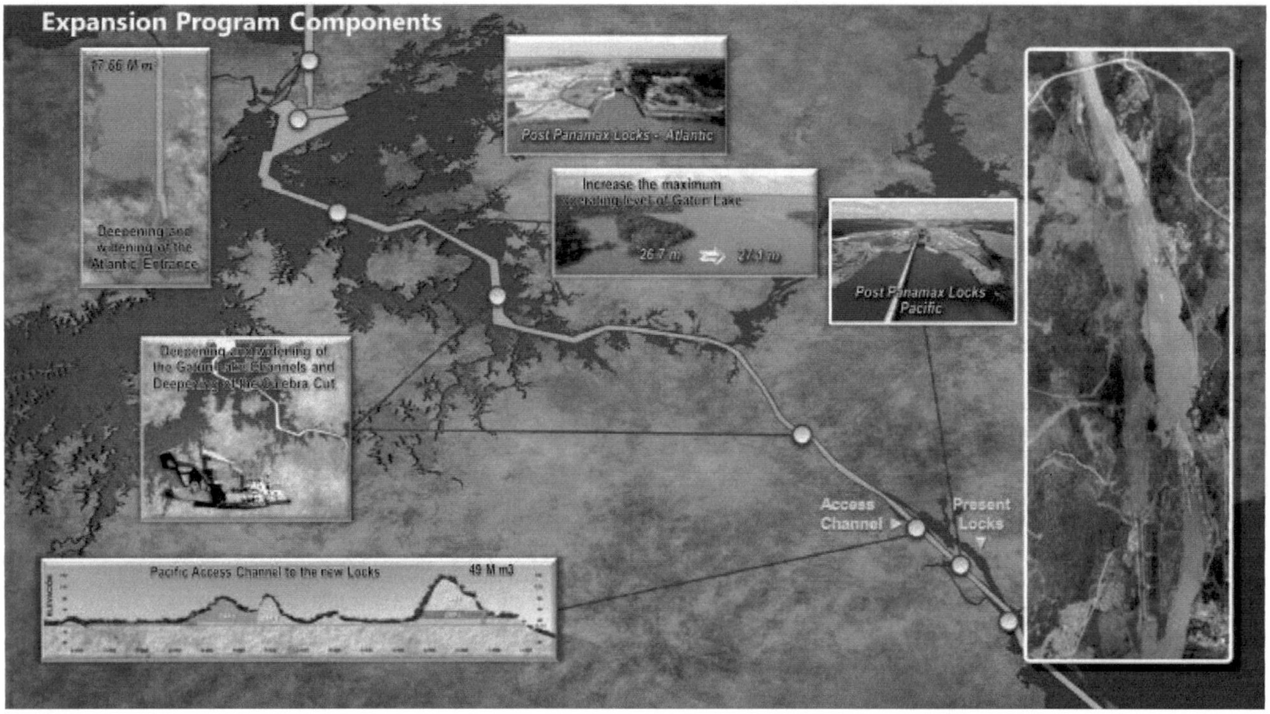

Figure 6: Seven Elements of Panama Canal Expansion Program

Figure 6 provides a graphic representation of various projects under the full PCEP. During the early integration of CH2M into the program, some of the major over-arching contract requirements, such as quality assurance and the program management information system, were established for the benefit of all seven elements of the PCEP. However, as time progressed and in alignment with the decision by ACP to execute the program management resources in an integrated team approach, a majority of the CH2M team focused on the largest program risk, the design-build Third Set of Locks Project (TSLP).

At its peak, CH2M reached a maximum staffing level of 40 supporting the ACP (with 150 staff assigned to the Third Set of Locks Project) in program management, construction management, risk management, quality assurance, project controls, document controls, correspondence, contract development, contingency planning, design oversight support and specialized support. With the successful implementation of the knowledge transfer strategy, that number stands at seven full-time CH2M/Jacobs staff today (still supporting 40 remaining ACP staff), whose primary focus is on post-construction activities such as engineering, quality assurance, mentoring as well as management and claims consultation.

PCEP Stakeholders

Four primary stakeholders are driving the PCEP:

People of Panama: Ultimately, the people of Panama are the primary stakeholders. They are the ones who gave approval to proceed and now benefit from the success. The Panama Canal is not only an immense source of pride and identity for Panamanians, it also is a major contributor of employment opportunities for the populace. In addition, the Panama Canal generates significant revenue that is distributed annually to the government, providing sustainable funding for the country's economic growth. The success and well-being of the Canal and its expansion is vital to Panama's economic future.

Autoridad del Canal de Panamá (ACP): The autonomous State-owned entity responsible for the operation and management of the Panama Canal. The ACP has financial autonomy, as well as ownership of the canal's assets. ACP has taken great pride in the successful administration of the canal for nearly 20 years and has a strong orientation toward self-performance.

Board of Directors: The eleven-member group responsible for establishing the Canal's policies for operation and management pursuant to the National Constitution, the Panama Canal Authority Organic Law and ACP Regulations.

Global Maritime Industry: The Canal's ultimate customers who will benefit from increased loadings, larger vessels, reduced canal transit times and lower transportation costs. The shippers expect reliability and efficiency of Canal operations for the safe and timely transit of their vessels. The impacts on the PCEP are already being felt as ports on both U.S. coasts rush to modernize their facilities to take advantage of the larger, neo-Panamax cargo ships that pass through the new locks.

One Team One Mission Organization

During the Canal transition from the American Panama Canal Commission (PCC) to the Autoridad del Canal de Panamá (ACP), the slogan "One Team, One Mission" referred to the coordinated and harmonious work accomplished between 1979 to 1999.

Expressing the meaning of the slogan at that time, a PCC employee said:

> We all attended the work meetings with the same spirit of cooperation, without egoism or arrogance…. We are all aware that we are part of the same objective and that the goal is not to reach December 31, 1999, but to proceed without showing that there has been a transfer.[6]

Within the PCEP, the slogan was revitalized as an attempt to overcome the struggles within the collaborative relationship between ACP and CH2M. By adopting the Integrated Program Management Model with ACP maintaining complete control of the program and CH2M serving in an advisory, mentoring role, the slogan once again emphasized the importance of teamwork in a collaborative effort that was intensified with a prescribed one-on-one structure for critical roles.

In a recent monograph on *Chaperoning* and practices of collaboration in the PCEP, Dr. Smits opined:

> The relation between ACP and CH2M Hill is more complex than what the slogan "One Team, One Mission" suggests or as the Invitation to Bid prescribed. In fact, these messages contradicted each other: the slogan refers to an equal relationship while the Invitation to Bid states that ACP is superior to CH2M Hill. Given the ambiguity of these formal relations, it is not surprising that project participants struggle to find satisfactory conditions for collaboration. [In the words of one ACP engineer in 2012]: *Somos la sombra de CH!* (We are the shadow of CH2M Hill). [7]

Within the institutional and professional mentorship of ACP's staff throughout the execution, this approach was cross-cutting from the program management development of tools and processes to establish the governance of the PCEP to a mirror manager approach of CH2M, with ACP counterparts, offering guidance, advice, and support for ACP's decisions.

As recognized by Dr. Smits in 2012: "In the transition period [1979-1999], role segmentation existed between the American Government as a mentor and the Panamanian protégé. Back then, the local employees were subordinate to the foreign leaders, and with chaperoning, a similar relationship is created…. Today [2012], however, the collaboration is intensified with a "one-on-one" structure. ACP indicates to aim at an equal

[6] Smits, K. (2013).

[7] Smits & van Marrewijk (2012)

relationship between the Americans and the Panamanians, just like how the transition phase is remembered.[8]

ACP and CH2M Teaming Arrangement

To meet the ACP's expectations, the CH2M team worked towards the goal of a fully integrated teaming arrangement which had the CH2M team embedded within the client's organizational team. This was all geared towards the One Team One Mission objective. The CH2M team was fully embedded with the ACP's organizational team.

This required collaboration, calibration and hard work as the ACP and CH2M groups came to the organization with very diverse and different backgrounds, cultures and disciplines, and formed into one team and to work as a disciplined unit. Partnership with outside entities was not a normal cultural practice for the client. The prerequisite ACP considered for CH2M personnel was a minimum of 20 years' experience in their area of expertise and prior work experience on a mega-program (>$1B). This proved to be an invaluable decision by ACP to bring to the PCEP CH2M personnel who had decades of experience dealing with challenging situations that might be encountered and then had the ability to pass along this experience by mentoring ACP counterparts.

The initial step to achieve a "One Team One Mission" team was an intensive team training event occurring off-site with key management members prior to starting the Locks project. Physical training events, team competitions as well as integration of cultural themes (outcome was to benefit an indigenous Indian tribe) helped to break down barriers and assist in building a cohesive team arrangement. Such teaming sessions had been used within ACP before PCEP and continue today under the current CEO Jorge Quijano.

With the extensive scope of the program it was essential for the program and, later the Locks Project, to have an integrated organization along all key roles and functional areas. This required not just complete integration but the ability to work vertically and horizontally within the client's organization, so as to work effectively and harmoniously. Everything from project controls to safety, the environment, risk, design, contracts, claims, quality, and construction management - all needed to work smoothly just like the water rising and lowering in the Locks Chambers with minimal turbulence to the massive ships passing across the Canal.

Communication is the Key

As noted above, the integrated organization was cross-cutting both vertically and horizontally. As the team was large and physically spread out, this was purposely addressed with integration occurring at all critical physical locations such as the Atlantic and Pacific construction sites and the ACP central office locations. This occurred amongst different program functions including quality, project controls, risk, safety, and further covered at the TSLP functions of design, construction management, resident engineering, quality assurance, claims and contracts, document controls, etc.

CH2M's internal publication, *The Little Yellow Book*[9], written by co-founder Jim Howland to immortalize the company's core values of respect and integrity, has been used for decades amongst its employees and with clients. This was shared by many CH2M staff with other members of the integrated team to reiterate such simple concepts such as "No matter what the organizational structure, if the people in it want it to work, it will."[10] Even in a difficult and complex program, this was preached.

[8] Ibid.

[9] Howland (1982).

[10] Ibid.

Communications were essential amongst the program team for success. This meant meetings at all levels strived to include attendance of ACP and CH2M, communications and meetings with the contractor were jointly held and leverage of decisions made by ACP were assisted by the knowledge of the integrated team being well informed and advised appropriately in consideration of moving the project forward. To achieve the goal of "One Team One Mission," communications with the shared knowledge of critical events and activities allowed timely and sound decisions. In following this process, ACP's communications extended to all stakeholders and was openly demonstrated with the public being allowed to visit the construction progress prior to the flooding of the new Locks in 2015.

Beyond the ACP / CH2M Team

The One Team One Mission did not only pertain to ACP and CH2M. As with any type of large program, the core values of integration, communication and common objectives were set to be inclusive of other entities supporting ACP's program execution. This included consultants, technical experts, and other professionals all sharing the vision for the expansion to become reality. While their scope was focused in the areas of support, the shared communications and drive towards acting as one unit was overarching.

In addition, the interfaces with the Contractor were conceived, again, towards obtaining the goal of completing the project. As challenges were faced, the ACP / CH2M Program team looked towards resolving situations timely and relied upon the contractual basis for support. Some hurdles were very complex as they involved time and financial impacts. All parties understood that, for the work to be completed, all parties would be relied upon to accomplish the goal. ACP decisions were geared towards achieving the ultimate goal - completing the Panama Canal Expansion Program successfully.

Mirror Management Counterpart Teaming

The organization was formed with the concept of 'mirror management' and 'counterpart teaming' (See Figure 7). Many of the key program organizational management positions had a mirror counterpart. That mirror counterpart was a CH2M partner in a teaming arrangement with the ACP counterpart. It was always clear to the Contractor (and CH2M) that the ACP partner was the contractually empowered entity and had full authority. However, the CH2M mirror manager offered guidance, advise, support, etc. to assist ACP's decision making and management of the works. This created an ability to mentor senior level talent, experience and background in the pursuit of the best possible outcome.

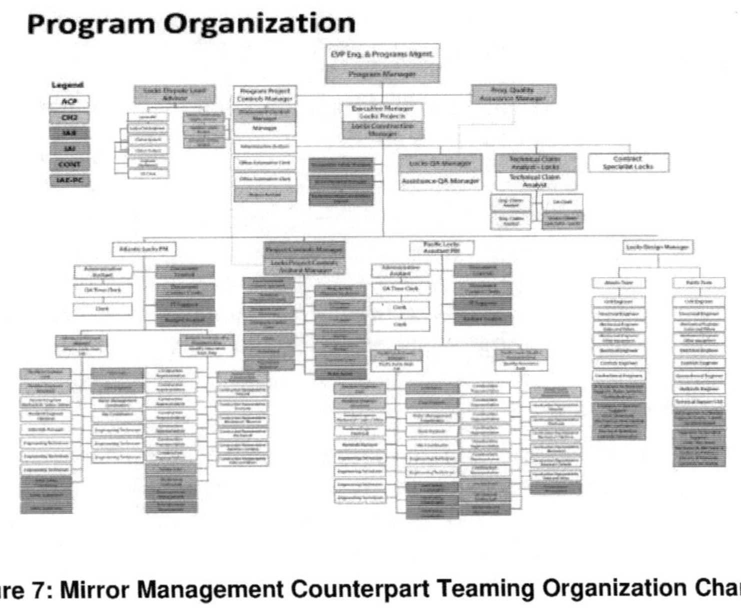

Figure 7: Mirror Management Counterpart Teaming Organization Chart

The objective was for ACP to take over the management role solely and for CH2M to work themselves out of the counterpart role which was achieved as shown in Figure 8 where only two of eight CH2M positions remained as primary roles by 2013. As part of One Team One Mission, the partner roles also had the benefit of the differences in perspectives, cultures and experiences with the common objective of moving the program forward.

The combined 2010 and 2013 organizational charts in Figure 8 demonstrate the concept of the mirror management of the ACP and CH2M partners early in the program (2010) and later (2013) with various ACP managers having taken over the management roles.

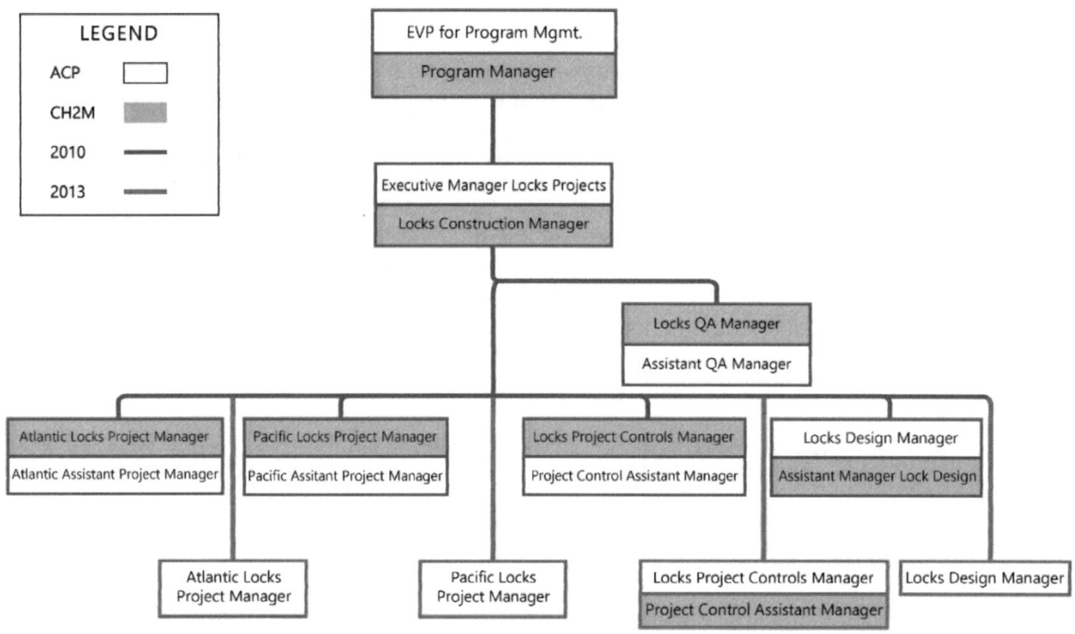

Figure 8: Comparison of 2010 and 2013 Organization Charts

During execution of the program, there was a continuous drive by senior management of ACP and CH2M to adhere to the one team concept of execution to leverage the advice, strengths and experiences of team members, as appropriate. The ultimate decisions were made by ACP but the organizational set-up was all pushing for the same objective - to succeed in the Expansion efforts with the best outcome for the new facilities in a timely manner and with responsible financial management.

This required different staff of both entities to recognize that, over time, their individual roles on the program/project would result in an element of 'musical chairs.' For example, the Figure 8 excerpt of the Organization Charts for the years of 2010 and 2013, respectively, clearly depicts that, over that 2.5-year period, the CH2M mentors soon either completed their project roles and demobilized or were moved into a subsidiary position to their ACP counterparts just as ACP staff were elevated to more senior roles in the organization.

The mentorship did not stop at the level of the mirror management roles. It also extended to the junior staff of ACP. ACP's organization already had a strong standard of continued education and training amongst its employees. However, with the execution of a large program and the mix of experienced team members, the continuous mentorship on various program and industry practices occurred. This included areas such as safety, quality, document controls systems, management, etc. Many of these continued to provide technology transfer or experiences, processes and tools to enhance the team and grow ACP staff. Thereby also fulfilling the role of CH2M working themselves out of particular roles to grow ACP overall.

Reach Back

The teaming arrangement between ACP and CH2M also allowed for leverage of CH2M resources on a reach back basis. This was utilized periodically for limited short-term assistance of experienced professionals and assisted in bringing new resources to the program team. New resources were needed as departures are unavoidable on a large program but also were used to cover new positions required, on a part-time or full-time basis as the program evolved during execution of the works.

Growth of a Nation

One of the outcomes of the One Team One Mission teaming arrangement and mentorship is the ensuing success of individuals associated with the Expansion program for the Panama professions and for ACP. Many of the mentored ACP staff are today moving forward with their careers based on their shared experiences and many are now working within the greater Panama professional engineering and construction community.

Design-Build Contract Development

The strategy for developing the PCEP was tailored to meet ACP's administrative requirements and the needs of the various construction projects that make up the overall program. The scope of work for many projects was focused on dry excavation and dredging, both of which were familiar to ACP. These projects, many of which were self-performed by ACP's existing maintenance crews, were treated in a relatively routine and conventional fashion, typically on a unit-price basis.

However, the **Third Set of Locks Project (TSLP**), by far the largest single capital works construction project performed in Panama (since the original Panama Canal construction) and was something entirely new. The project also needed to be integrated with the overall program. The project incorporated design concepts from various facilities around the world and no single installation used all the elements planned for the Panama Canal. Also, the sheer scale differentiated the project from the other locks, so a different approach was needed.

The ACP commissioned a detailed evaluation of applicable contract strategies and ultimately selected a design-build approach with the design and construction to be performed by a "world-class" consortium. The ACP felt this was the best way to manage risk, increase flexibility and innovation, bring appropriate expertise to the project and place full delivery responsibility with one entity. The requirements for the locks would still be controlled by ACP but the choice of how to deliver to the requirements would be owned by the design-build contractor.

Third Set of Locks Selection Process

The prequalification and selection of the design-build contractor for the Third Set of Locks contract was a process that spanned 23 months and involved several steps. The first was issuing a series of informational communications to the design and construction industry regarding the anticipated project concept, timeframe and other details. These communications were intended to generate awareness and interest across the industry worldwide.

Next, a Request for Qualifications (RFQ) was published with the intent to pre-qualify up to four tenderers which contained various experience and financial requirements. Although a joint venture or consortium was not specifically required, it was generally understood that the requirements of the RFQ would not likely be satisfied by any one individual company. The submitted Statement of Qualifications (SOQ) packages were evaluated and scored by the ACP and four separate teams were pre-qualified and invited to submit tenders for the project.

During the SOQ submission and evaluation period, the ACP/CH2M team continued preparation of the project Request for Proposals (RFP). The initial RFP was issued shortly after the pre-qualification process results were announced.

During the tender preparation period, a series of meetings was held with the tenderers to promote the exchange of ideas about the project's scope and the composition of the ACP's tender package. Prompted by comments and concerns raised by the tenderers, several refinements to the RFP evolved during this process and were incorporated into the tender documents. The process also gave the parties a chance to get to know each other and for the ACP to better judge the tenderers' concerns about subjects such as risk tolerance, financial capacity and dispute resolution expectations.

Risk Management

A key element of any Program Management model is risk allocation. For design and construction projects that are executed in traditional ways, the methods of assigning risk between an owner and contractors are well established. The basic principles are that risk should be assigned to one of the following:

- The party best able to manage risk
- The party best able to mitigate or minimize risk
- The party best able to overcome the adverse effects of actualized risk

Risk transfer works best when you get as close as possible to a "win-win" scenario. A risk profile with an imbalance inevitably leads to the detriment of one or more parties. Such analysis lends itself well to game theory and other analytical tools. The analysis of the risk profile itself also can be performed through probability analyses and simulation techniques.

Several risk allocation approaches can enhance chances for successful partnering:

- To the extent practicable, identify risks clearly and completely. Risks that are well understood can be readily quantified and will generally facilitate negotiations.
- Establish fair and reasonable provisions for relief of performance, cost or schedule commitments contained in given risk allocations. Fair and reasonable provisions for accommodating "uncontrollable circumstances" with appropriate risk allocation to both parties are more likely to be satisfactory than if contract provisions are viewed to be heavy-handed and one-sided.
- Recognize, through risk allocation and pricing provisions, that risk has an economic impact. For example, risk acceptance may require additional compensation. In so doing, avoid assigning risks to parties that do not have adequate power to control or manage the risk.

In the case of the PCEP, the integrated ACP-CH2M organization identified the risks and responsibilities during the tendering process and risk allocation was improved as a result of open dialogue. This risk and responsibility approach was maintained throughout the execution.

The relevance and importance of risk management can be generalized for future large infrastructure initiatives. For the Panama Canal Expansion Project, there were two main risk categories:

- Low-Bid Contracting and Delivery Risk; and
- Reputational Risk

With an eye toward a potential fourth set of locks program, it is interesting to note that the original canal efforts spanned more than 30 years, primarily because of a 20-year abandonment of the project between the French initiation and US completion, at a 2014 cost equivalent of almost US $17 billion. In contrast, the Third Set of Locks project and overall canal expansion was completed within a consecutive nine years at a final cost which is expected to be significantly below the equivalent cost of the original canal construction.

Request for Proposals

The RFP was generally based on the International Federation of Consulting Engineers (FIDIC) Yellow Book – a widely accepted international form of contract – since it was not known where the tenderers would be based and because it includes standard construction terminology well known in the industry. ACP requested permission and modified some clauses to add certain ACP requirements. As it turned out, the tenderers included companies from Europe, North America and Asia. It was decided to have a single contract for both the Atlantic and Pacific Locks to ensure consistency, maximize efficiencies and increase coordination.

A significant effort was made to find an appropriate balance between performance-oriented and prescriptive employer's requirements. Although the ACP wanted the design-build contractor to have the greatest flexibility possible, certain requirements were necessary to meet specific concerns and needs. This included having the new locks facilities interface properly with existing, ongoing operations.

Ultimately the RFP incorporated updates of the tenderer stipend amount, several risk-sharing provisions, allocation of responsibilities, extensive information regarding the existing geotechnical site conditions, key construction materials escalation protection and an advance payment scheme to help the design-build contractor with early cash flow. All these enhancements were designed to minimize uncertainty for the tenderers with the critical goal of reducing the tender price and ensuring the Third Set of Locks project would be able to progress without delay, and that it would be completed within the required budget parameters.

The tender process was based on a fixed-price contract with selection based on best value, considering both technical merit and cost. Tender packages were evaluated and technical considerations were scored by the ACP. Subsequently, separate cost proposals were opened, and the successful tenderer was determined by the ACP based on the numerical combination of the technical score and cost, according to the formula outlined in the RFP. The ACP team led the proposal and tendering process through its procurement procedures, as modified for the PCEP.

Tendering Process Results

The tendering process resulted in the ACP receiving three qualified tenders with a wide range of tender prices. The lowest price, which was submitted by Grupo Unidos por el Canal (GUPC), was a few percentage points below the employer's estimate and roughly $1 billion below the next lowest tender. ACP also gave GUPC the highest technical score. The tender process accomplished its goal of receiving a qualified tender at a price that was within the established estimate, but the relatively low bid and large range in tender prices was an indication that difficulties might lie ahead.

Achieving Collaboration Among Diverse Cultures

The Next Generation Infrastructures Foundation of The Netherlands funded a four-year research project by Karen Smits on collaboration practices in the PCEP. This research resulted in the 2013 publication previously referenced in footnote 6 above. Dr. Smits' year-long research as an embedded PCEP researcher resulted, among other findings, in the following ***Collabyrinth***[11] depiction of the manifest and concealed practices which can ultimately enhance or hinder the collaborative practices in the everyday organizational life of project participants.

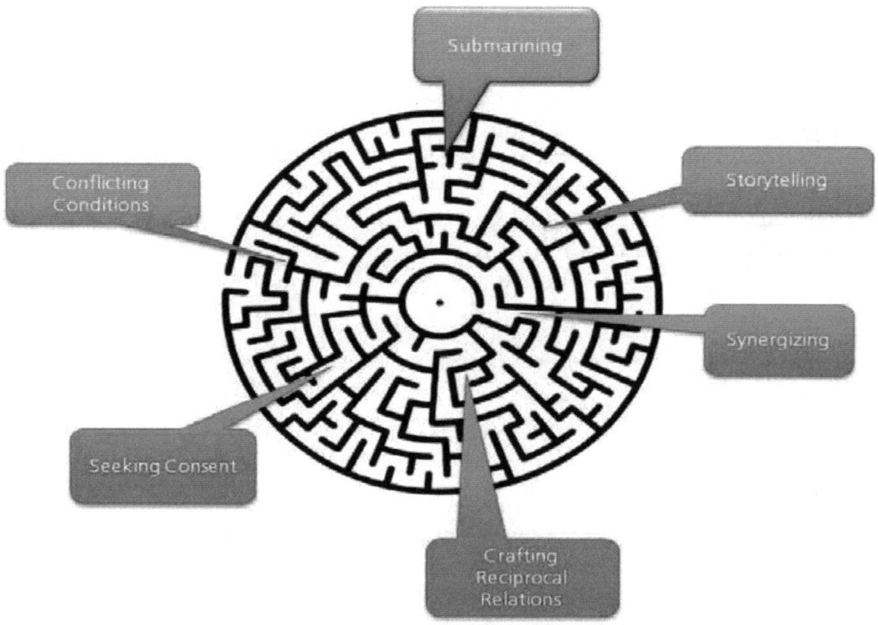

Figure 9: *Collabyrinth (neologism of Collaboration and Labyrinth)* as presented by Karen Smits, Ph.D.

In the PCEP relationship between ACP and CH2M both manifest and concealed practices consisting of interconnected actions and activities are illustrative of how different project participants make sense of the cultural complexities in their everyday work environment.

By observing the collaborative efforts of both parties to the integrated program management model implemented for PCEP, Dr. Smits cited examples which exemplified elements of the scholarly literature on organizational management. For example,

- The partners struggled over the different interpretations about the role that CH2M consultants would play in delivering program management services. CH2M was guided by its traditional independent delivery role while ACP envisioned a purely advisory role of training, monitoring and mentoring. The parties soon became aware of these routine practices and changed toward a more discursive mode suitable for project enhancement, confirming the scholarly literature.[12]
- Conflicts over goals are present in everyday organization practice and are, essentially, an evident part of collaborative projects.[13]
- A great sense of pride dominated the project narratives told by ACP and CH2M employees resulting in a more profound relationship with the project again confirming the scholarly literature.[14]

[11] Smits, K. (2013)

[12] Smits (2013) but see also Bresnen & Marshall (2011), Engwall (2003) and Van Marrewijk et.al. (2008)

[13] Smits (2013) but see also Gherardi (2000), Nicolini et.al. (2003) and Vaaland (2004).

[14] Smits (2013) but see also Veenswijk and Berendse (2008).

Conversely, GUPC's collaboration was rather loose as project participants saw themselves as only being in a temporary state of working together. Additionally, the distance between the Atlantic and Pacific project sites (approx. 60 kms) and the linguistic differences between the two sites created a scenario where competition between the two site offices was often stronger than collaboration. This situation provided a ripe environment for CH2M to create collaborative opportunities between the parties (and the subcontractors) to the TSLP Design/Build contract.

Although the construction sector is not unique in the elements of risk and uncertainty in the workface, the key driver for collaboration in megaprojects is the need to surmount the lack of competencies and scarce resources to create value together while crafting reciprocal relations of mutual trust and respect.[15] Tellingly, Dr. Smits concluded that "collaboration advanced due to a strong personal drive to finalize the project successfully rather than because the project organization stimulated such a work environment."[16]

In a recent paper presented on behalf of ACP at the Dubai International Project Management Forum what differentiates the Canal Expansion project from most others is, without a doubt, the diversity of labor. Over 41,000 jobs were generated and, of those, over 37,000 were Panamanian workers who contributed their efforts and dedication to make it possible, in contrast to what occurred with the construction of the original Canal.[17]

To this number, several thousand were a foreign labor pool of men and women coming from 80 different parts of the world including: Spain, Portugal, Colombia, Italy, Venezuela, Chile, Romania, Nicaragua, United States, Mexico, Belgium, Costa Rica, Moldova, Argentina, Holland, Peru, Dominican Republic, Great Britain, Honduras, Hungary, Brazil, Ecuador, Poland, Uruguay, El Salvador, South Korea, Ireland, Philippines, Cuba, South Africa, Germany, Ukraine, Canada, Russia, Albania, Angola, Bosnia and Herzegovina, China, France, Guatemala, North Korea, Paraguay, Sweden, British Indian Ocean Territory, Bulgaria, Croatia, Slovenia, Turkey, Vatican City, Bahamas, Bolivia, Eritrea, Haiti, India and Iceland.

This cultural diversity represented a new challenge. All cultures, styles and personalities united for a common purpose. The goal was fulfilled because early on, each one assumed, the Panama Canal Expansion as their own work.[18]

Locks Project Claims and Disputes

The locks project dispute process detailed in the contract begins with a notice of claim and subsequent details of claim to be provided by the contractor. The employer responds to the claim with an approval/disapproval followed by a determination. At that point, if there is no agreement, either party can refer the issue to the project Dispute Adjudication Board (DAB), which has been established jointly by the contractor and the employer to rule on referred disputes. After a DAB ruling is issued, if either party is dissatisfied, it may escalate settlement of the dispute to be decided by International Chamber of Commerce (ICC) arbitration.

During the seven-year life of the TSLP contract, significant numbers of claims were lodged by the contractor. Both the contractor and the owner established teams dedicated to the preparation and response to these

[15] Smits (2013) but see also Sackmann & Friesl (2007) and Vangen & Huxham (2003).

[16] Smits, K. (2013).

[17] Marotta, I. (2016).

[18] Ibid.

claims. Ultimately, these numerous contractor claims became specific DAB referrals. Those that have not been agreed by both parties, as well as several employer counterclaims, are currently before the ICC arbitration tribunals in separate distinct multimillion dollar cases between the parties.

Grand Opening of the Panama Canal Expansion Project in June 2016

In June 2016, all stakeholders celebrated the opening of the Third Set of Locks as well as the start of a three-year maintenance contract which extends through June 2019.

At the time, the CEO of the Autoridad del Canal de Panamá stated that CH2M had been a trusted partner since the PCEP began:

> *Their proven experience in the international program management arena has added tremendous value to our integrated team. Providing professional support at all levels of their organization, CH2M worked with the Canal Authority to assist in achieving successful completion.*

Figure 10 Grand Opening of the Panama Canal Expansion Project in June 2016

Megaproject Lessons for Future Infrastructure Initiatives

As is normal on programs of this size and scope, there are several lessons that can be generalized for future large infrastructure initiatives. For the Panama Canal, these fall into three main categories.

Low-Bid Contracting and Delivery Risk: Interdependence has become the norm in the global economy. The Third Set of Locks Project has required the planning, design and construction expertise of many firms from around the world. To that end, the ACP adopted a Program Management model that promised ownership control over the PCEP while balancing the risk appropriately to the design-build contractor. For the selection of the contractor, the ACP utilized a Non-Negotiated tender process based on Best Value for the Third Set of Locks Design-Build Project. However, with the significantly lower bid, as compared to the other bidders, the construction risk was essentially transferred back to ACP, which has gone to extraordinary lengths with the contractor to enable the successful completion of the work.

To date, the progress of the PCEP has followed the path of many "mega-programs" with respect to cost and schedule overruns. The large, complex, inherent nature of these programs make them subject to a wide array of challenges. Each program is unique and the challenges associated with each also are unique. As with any major program, alternate scenarios to complete the work must be kept intact – this too is a risk management measure that is necessary no matter what program management model is in place.

Reputational Risk: In today's world of instant communications, the need for fast-paced understanding of situations and potential resolution is paramount. Cultural incompatibilities must be identified and, in some cases, may actually present a fatal flaw to a program's ultimate success. Contract terms and conditions are a tool and cannot by themselves be expected to resolve overwhelming differences. More than ever, the understanding of a win-win relationship is critical.

Cultures, Customs and Languages: On major programs – particularly those with a multi-nation delivery platform – learning and respecting the political, cultural and economic environment of the host country is essential to establishing a trusted relationship with the owner and all stakeholders. In the case of the PCEP, beyond absorbing and understanding Panama's social and political culture, there is an added challenge of collaborating with people from around the world all bringing with them their particular values, work ethic, customs and languages.

The invaluable aspect of Program Management to the PCEP is the management of the interrelated portfolio of projects and execution of work with ACP senior managers approving and signing off on all plans, incorporating the applicable processes and procedures. While English was the contract language, it was not the primary language for all contractors which posed some challenges in practice.

For that reason, special care and attention should be taken to ensure all parties involved clearly understand actions and intent – colloquialisms or slang must be avoided. And for the CH2M team, though all program and project plans, documentation and reporting are conducted in English, fluency in Spanish for at least part of the primary program staff assisted in developing stronger bonds with the ACP. The overall integrated program efficiency was not only increased by providing staff who were proficient in the local language but also those with past international working experience or the ability to easily adapt to a different language and culture. With any of the program models, communications issues and the ability to adapt culturally – not necessarily technical challenges – can lead to the most substantial program risks.

Considerations for Future Infrastructure: The PCEP confirms that today's program management teams need to go well beyond providing technical expertise. Working hand-in-hand with stakeholders and peers, an integrated team can mitigate some megaproject lessons:

- People, processes and tools … in that order
- Early involvement of program management team
- Team integration and constant communication is crucial
- Total transparency in bidding, evaluation and contract award
- Clear and precise contractor agreements about roles and responsibilities
- Consortia with many diverse members present considerable challenges
- Trusted advisors managing change, complexity and unpredictability
- Business managers balancing cost and schedule requirements with the terms of the contract
- Safety advocates, making sure that workers are returned to their families safely at the end of the work day
- Human resource counselors, creating a positive work environment
- Politicians, considering all stakeholder perspectives and building consensus
- AND technical experts, bringing engineering best practices and solutions to the challenge at hand.

References:

Bresnen, M., & Marshall, N. (2011). Projects and Partnerships: institutional processes and emergent practices. In P. W. G. Morris, J. K. Pinto & J. Söderlund (Eds.), *The Oxford Handbook of Project Management* (pages 154-174). Oxford University Press.

Engwall, M. (2003). No project is an island: linking projects to history and context. *Research Policy,* 32 (pages 789-808).

Gherardi, S. (2000). Practice-Based Theorizing on Learning and Knowing in Organizations. *Organization* 7 (pages 211-223).

Howland, J. (1982). *The Little Yellow Book.* CH2M HILL Consulting Engineers.

Marotta, I. (2016). *Mega Projects Management.* Dubai International Project Management Forum 2016.

Marotta, I. (2017). *Lecciones Aprendidas del Programa de Ampliación del Canal*, VIII Congreso Internacional de Dirección de Proyectos, PMI Panamá.

Nicolini, D., Gherardi, S., & Yanow, D. (2003a). Introduction. In D. Nicolini, S. Gherardi & D. Yanow (Eds.), *Knowing in Organizations: A Practice-Based Approach (pages 3-31).* New York: M.E. Sharpe.

PMI's *Pulse of the Profession®* In-Depth Report: *The Impact of PMOs on Strategy Implementation (2013)*

PMI's *Pulse of the Profession®* In-Depth Report: *Capturing the Value of Project Management* (2015), page 3.

Sackmann, S. A., & Friesl, M. (2007). Exploring cultural impacts on knowledge sharing behaviour in project teams - results from a simulation study. *Journal of Knowledge Management,* 11(6) pages 142-146.

Smits, K. (2013). *Cross Culture Work: Practices of Collaboration in the Panama Canal Expansion Program.* Next Generation Infrastructures Foundation. ISBN/EAN 978-90-79787-49-4; pages 220-258.

Smits, K. & van Marrewijk, A. Chaperoning: practices of collaboration in the Panama Canal Expansion Program. *International Journal of Managing Projects in Business,* 5 (3) pages 449-451.

Vaaland, T. I. (2004). Improving project collaboration: start with the conflicts. *International Journal of Project Management,* 22 (pages 447-454).

Van Marrewijk, A., Clegg, S., Pitsis, T., & Veenswijk, M. (2008). Managing public private megaprojects: Paradoxes, complexity, and project design. *International Journal of Project Management,* 26 (pages 591-600).

Vangen, S., & Huxham, C. (2003). Enacting Leadership for Collaborative Advantage: Dilemmas of Ideology and Pragmatism in the Activities of Partnership Managers. *British Journal of Management,* 14 (pages S61-S76).

Veenswijk, M., & Berendse, M. (2008). Constructing new working practices through project narratives. *International Journal Project Organisation and Management,* 1(1) pages 65-85.

34th PIANC WORLD CONGRESS PANAMA 2018 May 7 to 11

CANAL DE PANAMÁ

FINANCING OF THE PANAMA CANAL EXPANSION

Eida Gabriela Saiz

Treasury and Finance Manager

Panama Canal Authority

34th PIANC WORLD CONGRESS PANAMA 2018 May 7 to 11

CANAL DE PANAMÁ

AGENDA

- Objectives
- Bankable Proposal
- Financing Sources
- Negotiation
- Learned Tips

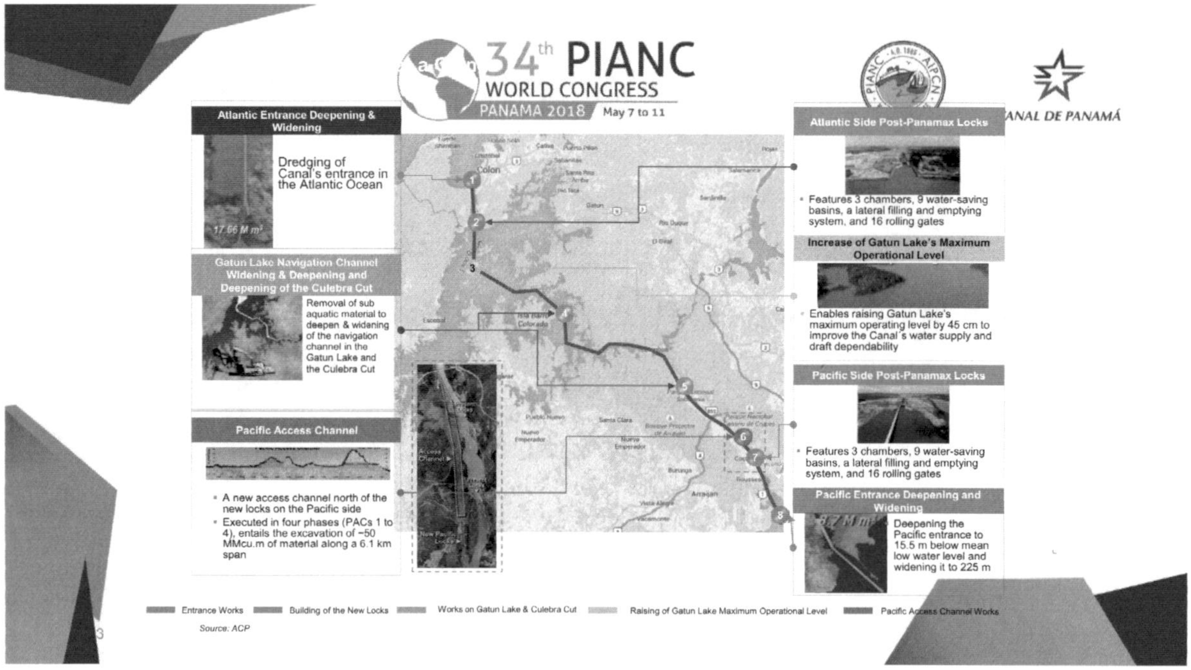

34th PIANC WORLD CONGRESS
PANAMA 2018 May 7 to 11

CANAL DE PANAMÁ

Financing objectives

- Expansion Program financial needs - up to US$2,300 Million
- Tenor - longest possible
- Pricing - as per financial strength of the ACP
- Drawdrown – flexibility
- Prepayment - allowed
- No guarantees
- Untied financing - flexibility of suppliers
- Global diversity of lenders

34th PIANC WORLD CONGRESS PANAMA 2018 May 7 to 11

CANAL DE PANAMÁ

Bankable Proposal

- Strength of Panamanian administration with the legal framework, ability to manage operation, pricing
- Develop strong financial model (inflows/outflows) includes:
  - Demand model: future demand for over an 18 year horizon
  - Capacity model: types of vessels, segments, draft beams that transit Panama Canal.
  - Risk model: projected execution drawdown, costs and contingencies
- Independent technical studies/ peer reviews on social and environmental condition, insurance and legal framework.
- Hired competitive world class legal and finance advisors

34th PIANC WORLD CONGRESS PANAMA 2018 May 7 to 11

CANAL DE PANAMÁ

Market Sounding Results

Financing Sources

	Source	Type	Potential Amount (Up to)	Tenor (years)	Covenants
Loan Market	Structured / Finance	Syndicated bank loan w/ credit enhancements based on cash flow from the expansion/ current operations	$1.5 billion	5 to 7	Linked to Expansion Program
	Corporate	Direct bank loan to ACP, based corporate credit risk	$2.3 billion	Up to 7-8	Corporate governance related

Financing Sources

	Source	Type	Potential Amount (Up to)	Tenor (years)	Covenants
Multilateral	Direct Loan	Direct loan to ACP assessment and Project's risks	$400 million each	Up to 15	Complex Covenant Package (project + corporate)
ECA's	Export Credit/Untied	Direct credit to ACP	Depends on the source	Up to 20	

34th PIANC WORLD CONGRESS PANAMA 2018 May 7 to 11

CANAL DE PANAMÁ

Financing Sources

	Source	Type	Potential Amount (Up to)	Tenor (years)	Covenants
Capital Market	Private Placement	Small number of investors	$1 billion	Up to 30	Light Covenant Package
	Public/144A	Larger investor pool (public / qualified investors)			

Drawback - Negative Carry

34th PIANC WORLD CONGRESS PANAMA 2018 May 7 to 11

CANAL DE PANAMÁ

Common Agreement Structure

- Standard terms and conditions applying to all credit facility lenders.
- Specific terms of each agency. i.e. interest rates, prepayment notices, etc
- Intercreditor Agreement: Terms and conditions to address amendments, waivers, etc.

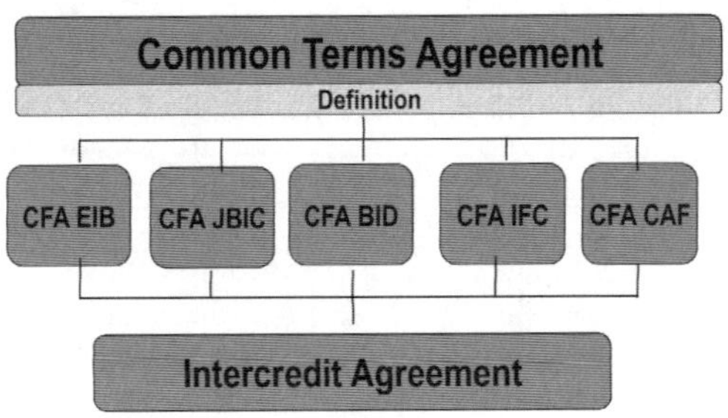

Confidencial

34th PIANC WORLD CONGRESS PANAMA 2018 May 7 to 11

CANAL DE PANAMÁ

Financing objectives

- Expansion Program financial needs - up to US$2,300 Million
- Tenor - longest possible
- Pricing - as per financial strength of the ACP
- Drawdrown – flexibility
- Prepayment - allowed
- No guarantees
- Untied financing - flexibility of suppliers
- Global diversity of lenders

✓ 5 Development agencies

✓ 20 years with 10 year grace period -amortization

✓ Spread according to A1 credit rating

✓ Disbursements from lowest cost

✓ Prepayment - allowed

✓ Sole Financial Strength of ACP

✓ Open to competition

✓ Agencies represent countries around the world

34th PIANC WORLD CONGRESS PANAMA 2018 May 7 to 11

CANAL DE PANAMÁ

Amount	**USD2,300 millions (44% of Expansion Program)**
Tenor	20 years, 10 years of grace period
Interest Rate	Libor 6M + applicable margin
Committment Fee	0.25% and 0.50% IFC
Amortization	May 15, 2019

Credit Facilities	**Amount (million)**	**Interest**
European Investment Bank (EIB)	100	Libor 6 m + 0.811%
European Investment Bank (EIB)	300	Libor 6 m + 0.824%
European Investment Bank (EIB)	100	5.196%
Banco Latinoamericano de Desarrollo (CAF)	150	Libor 6 m + 1.20%
Banco Latinoamericano de Desarrollo (CAF)	150	Libor 6 m + 1.40%
International Finance Corporation (IFC)	300	Libor 6 m + 1.30%
Interamerican Development Bank (IADB)	400	Libor 6 m + 1.05%
Japan Bank for International Cooperation (JBIC)	800	Libor 6 m + 0.75%
Total	**2,300**	

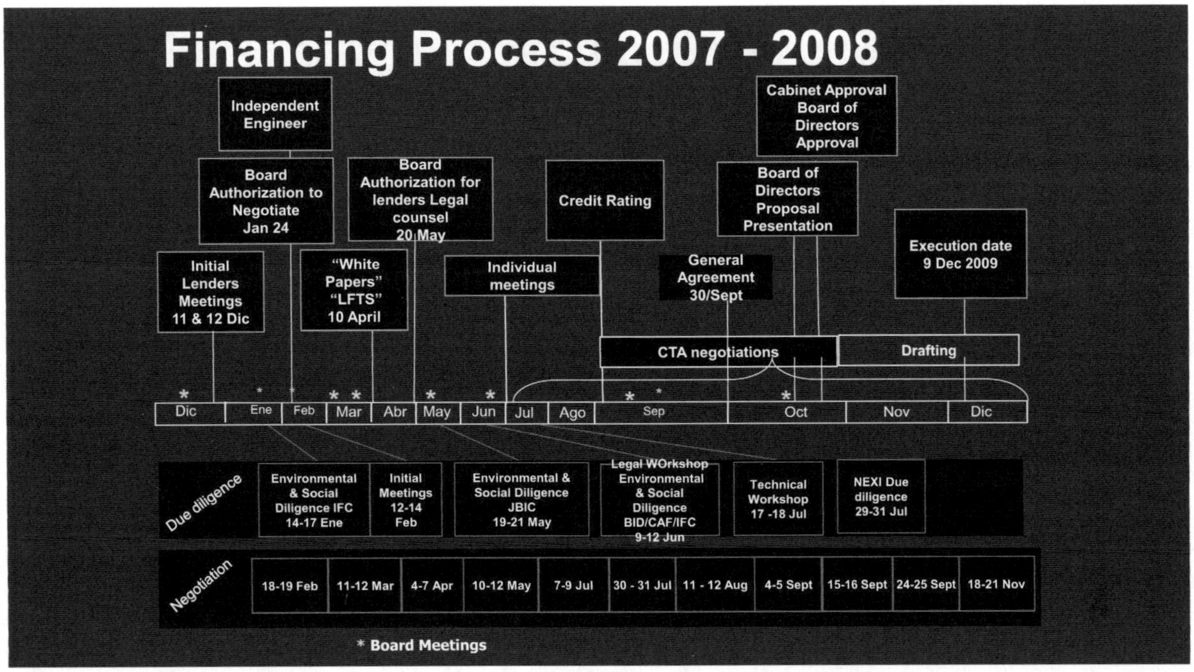

34th PIANC WORLD CONGRESS PANAMA 2018 May 7 to 11

CANAL DE PANAMÁ

- ACP coordinates one annual visit to all five lenders.

Benefits:

- All lenders have the same information at the same time
- Key ACP personnel participates
- Networking/shared opinions among lenders
- Site visits taking into consideration lenders inquiries.
- Different participants from lenders in every visit.

November 2010

December 2012

June 13, 2018 Confidencial 14

34th PIANC WORLD CONGRESS PANAMA 2018 May 7 to 11

CANAL DE PANAMÁ

ACP Debt Service Profile

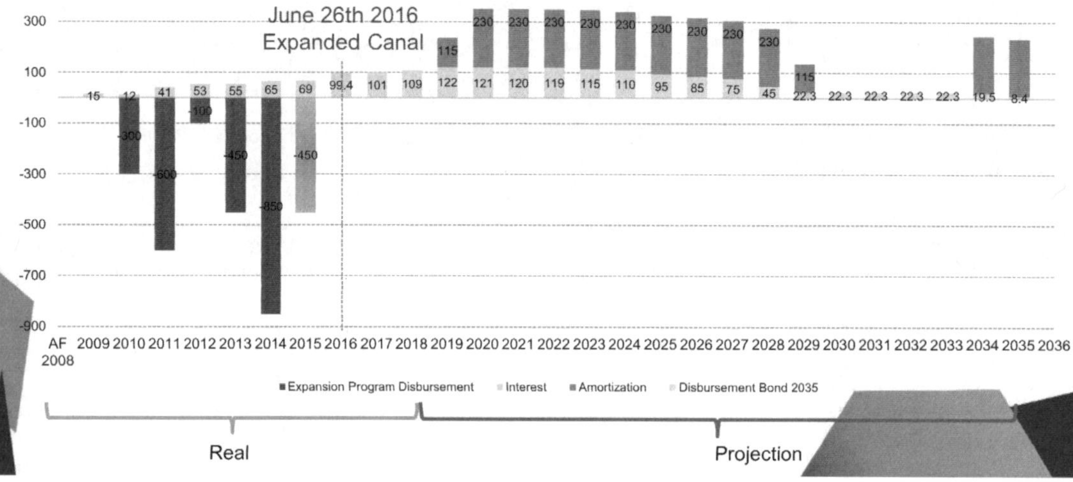

34th PIANC WORLD CONGRESS PANAMA 2018 May 7 to 11

CANAL DE PANAMÁ

Lessons learned

- Hire international financial and legal experts to establish long form term sheet according to ACP legal framework, Corporate Governance, Financial Strength
- Hire one law firm to represent all lenders as legal counterpart
- Hold common legal, technical and financial due diligence meetings.
- Individually reach agreements with potential lenders
- Mirror common term agreement obligations/covenants to contractors /subcontractors: Social and environmental clauses, Prohibited practices and ethics
- Use /adapt ACP existing reports to include lenders requirements/ aligned with national/other reporting requirements
- Agree on one annual meeting to inform lenders at the same time
- Common sharepoint for due diligence sharing information and to comply with CTA obligations
- Keep the Board of Directors abreast of the advances.
- Document evruthing during negotiation and execution of loan agreement, visits, meetings.
- Coordinate lenders internal auditors (development agencies)

AUTHOR INDEX

The World Association for Waterborne Transport Infrastructure (PIANC)
General Secretariat
Boulevard Roi Albert II, 20 – B. 3
B-1000 Brussels
Belgium

ISBN 978-1-5108-8705-3

CURRAN ASSOCIATES INC.
proceedings.com